The Future of Internal Combustion Engines

Edited by Antonio Paolo Carlucci

Published in London, United Kingdom

IntechOpen

Supporting open minds since 2005

The Future of Internal Combustion Engines
http://dx.doi.org/10.5772/intechopen.76548
Edited by Antonio Paolo Carlucci

Contributors
Mateos Kassa, Carrie Hall, Akii Ibhadode, Raphael Ebhojiaye, Selahaddin Akansu, Mehmet İlhan İlhak, Nafiz Kahraman, Selim Tangoz, Antonio Paolo Carlucci

Notice
Statements and opinions expressed in the chapters are these of the individual contributors and not necessarily those of the editors or publisher. No responsibility is accepted for the accuracy of information contained in the published chapters. The publisher assumes no responsibility for any damage or injury to persons or property arising out of the use of any materials, instructions, methods or ideas contained in the book.

First published in London, United Kingdom, 2019 by IntechOpen
IntechOpen is the global imprint of INTECHOPEN LIMITED, registered in England and Wales, registration number: 11086078, The Shard, 25th floor, 32 London Bridge Street
London, SE19SG – United Kingdom
Printed in Croatia

British Library Cataloguing-in-Publication Data
A catalogue record for this book is available from the British Library

Additional hard and PDF copies can be obtained from orders@intechopen.com

The Future of Internal Combustion Engines
Edited by Antonio Paolo Carlucci
p. cm.
Print ISBN 978-1-83881-930-9
Online ISBN 978-1-83881-931-6
eBook (PDF) ISBN 978-1-83881-932-3

We are IntechOpen,
the world's leading publisher of
Open Access books
Built by scientists, for scientists

4,300+
Open access books available

116,000+
International authors and editors

125M+
Downloads

Our authors are among the

151
Countries delivered to

Top 1%
most cited scientists

12.2%
Contributors from top 500 universities

Interested in publishing with us?
Contact book.department@intechopen.com

Numbers displayed above are based on latest data collected.
For more information visit www.intechopen.com

Meet the editor

Antonio Paolo Carlucci is Associate Professor in "Systems for Energy and Environment" in the Department of Engineering for Innovation at the University of Salento, Italy, where he is responsible for the internal combustion engine laboratories. He received his MS and PhD at the University of Lecce in 2000 and 2004 and also worked at the University of Illinois in Urbana-Champaign, Illinois (USA). His research interest is mainly in combustion strategies with alternative fuels. He has taken part in different national and international research projects. He is the author of three book chapters, about 40 international journal papers, and approximately 40 papers presented during international and national congresses. He is the editor for two international journals and guest editor for a special issue on dual-fuel combustion.

Contents

Preface

Based on previsions, the reciprocating internal combustion engine (ICE) will continue to be widely used in all sectors: transport, industry, and energy production. Therefore, ICE development, while complying with limitations of pollutants as well as CO_2 emission levels and maintaining or increasing performance, will certainly continue for the next few decades.

In the last three decades, significant effort has been made to reduce pollutant emission levels emitted by ICEs. More recently, attention has been given to CO_2 emission levels too, due to well-known problems concerning global warming and oil depletion.

It is widely recognized that one single technology will not completely solve the problem of CO_2 emissions in the atmosphere. Rather, the different technologies already available will have to be integrated, and new technologies developed, to obtain substantial CO_2 abatement.

In this book, several contributions have been collected describing some of the latest research development in ICEs: dual-fuel combustion, alternative fuels, and light-weight materials.

Antonio Paolo Carlucci
University of Salento,
Lecce, Italy

Section 1

The Challenges of Future Internal Combustion Engines

Chapter 1

Introductory Chapter: The Challenges of Future Internal Combustion Engines

Antonio Paolo Carlucci

1. Introduction

Based on previsions, reciprocating internal combustion engine (ICE) will continue to be widely used in all sectors: transportation (land, sea, and sky), industrial, and energy production. For example, despite the significant electrification of powertrains based on light and heavy duty ICE for land transportation—the sector with the highest penetration of electrification—thermal engine will remain the prime mover. In fact, a majority of the scenarios investigated consider that various types of powertrains using liquid or gaseous carbon-based fuels would still cover up to 80% of the world fleet in 2050 (even if the major part of these powertrains will certainly be hybridized and partly electrified) [1, 2]. Therefore, the ICE development, finalized—based on the application field—at complying with limitations of pollutants as well as CO_2 emission levels while maintaining or increasing performance, will for sure continue for the next decades.

In the last three decades, a significant effort has been done finalized at reducing pollutant emission levels emitted by ICE: this research and development activity has led to a wide portfolio of technological solutions like improved injection systems and air management, more accurate design of the whole combustion system, aftertreatment devices, and so on [1].

More recently, attention has been addressed to CO_2 emission levels too, due to well-known problems concerning global warming and oil depletion. Dealing with CO_2 reduction requires a systemic approach for the following reason: nowadays, the concept of fuel for reciprocating internal combustion engines must be expanded, given that the electrification of powertrain is leading the engines to use electricity, besides conventionals (fossil, bio, and alternative), as fuels. Therefore, CO_2 is emitted not only during the engine operation but also when the *fuel* (then including the electricity) is produced. As a consequence, CO_2 reduction is currently pursued not only during the engine operation, but also at fuel and electricity production levels. Moreover, the overall result of CO_2 reduction is expected to be reached in a contest where the energy demand is significantly increasing.

It is widely recognized that one single technology will not solve completely the problem of CO_2 emissions in the atmosphere. Rather, the different technologies already available will have to be integrated in order to solve this global problem. In the following, the technologies already available will be analyzed, at fuel production, electricity production, and engine design levels.

At **fuel production level**, biofuels—gaseous or liquid—represent a prosecutable solution in terms of CO_2 reduction, given the "virtually zero" CO_2 cycle characterizing them. A biofuel is a fuel produced through contemporary biological processes, such as agriculture and anaerobic digestion, while geological processes from prehistoric biological matter produce fossil fuels, such as coal and petroleum. If CO2 emissions can be lowered by biofuels, pollution problems remain substantially unchanged. However, respecting sustainability criteria for biofuels limits the exploitation of biomasses for biofuel production.

Other investigated solutions are synthetic fuels, as hydrogen, syngas, and the so-called e-fuels. Their production, which can contribute at recycling the CO2 released in the atmosphere during combustion, requires in several cases a significant energy amount. Therefore, synthetic fuels make sense only if the energy required to produce them is obtained from renewable sources.

A good midterm solution is recognized to be the utilization of alternative fossil fuels already available in nature like methane, main constituent of natural gas. In this way, the impact of the production phase of fuels is mitigated. The utilization of methane leads to the reduction, but does not avoid the emission of both CO2 and pollutants.

At **electricity production level**, a big contribution to CO2 reduction has been provided in the last years by the significant spreading of renewable energy source exploitations. Photovoltaic plants, wind farms, geothermal and hydro power plants, currently account for more than 20% of global energy demand. The big problem of these technologies is represented by their availability, intrinsically limited in time and space, as well as their reliability. It is evident, then, that currently they are considered as integrative—rather than substitutive—energy sources. A possible solution to the problems mentioned above would be to develop technologies able to store the electricity produced in excess during the periods of high availability (as already done nowadays in several hydroelectric power plants were electricity in excess, produced during night hours by thermal power plants, is used to pump water at high altitude, so recharging the hydroelectric power plants for the following day). In this view, the exploitation of the electricity produced in excess when available could be used to produce the abovementioned synthetic fuels, both gaseous and liquid.

Finally, at **engine design level**, a substantial contribution to CO2 reduction can derive from the abatement of auxiliary power losses, but also recovering the thermal power usually wasted at the exhaust, for example, through technologies like turbocharging and turbocompounding, organic Rankine cycles, and thermoelectric generators. Moreover, for several applications, like transportation sector, engine lightweighting as well as downsizing lead to the reduction of overall weight of the transportation vector. In the first case, the engine weight reduction is obtained using lighter materials, while in the second case, the goal is to reduce the engine displacement keeping unchanged the engine performance. Hybrid powertrains—where multiple forms of motive power are available—also lead to the reduction of CO2 levels, because they allow to operate the thermal engines in the areas of the characteristic map with the highest values of fuel conversion efficiency. If a storage system is integrated in the powertrain, then other technologies leading to CO2 reduction can be also applied, like kinetic energy recovery systems.

As previously said, all the above technologies cannot be considered as isolated, but they will all contribute to CO2 abatement. In order to reach this result, they will require an always increasing level of integration in the future. ICE development will be then finalized at providing a machine suitable to exploit all the opportunities for reducing CO2 emissions while satisfying the energy demand.

Introductory Chapter: The Challenges of Future Internal Combustion Engines
DOI: http://dx.doi.org/10.5772/intechopen.83755

Author details

Antonio Paolo Carlucci
University of Salento, Lecce, Italy

*Address all correspondence to: paolo.carlucci@unisalento.it

IntechOpen

References

[1] Ertrac. Future Light and Heavy Duty ICE Powertrain Technologies [Internet]. 2016. Available from: https://www. ertrac.org/uploads/documentsearch/ id42/2016-06-09_Future%20ICE_ Powertrain_Technologies_final.pdf

[2] Magna. The Road to Electrification— The Magna Powertrain Approach [Internet]. Available from: http:// www.drivetrainforum.com/media/ departments/communications/events_1/ dtf/2018/vortraege_1/DTF2018_ Lecture_MPT_Stephan_Weng.pdf

Section 2

CO_2 Reduction from Internal Combustion Engines

Chapter 2

Dual-Fuel Combustion

Mateos Kassa and Carrie Hall

Abstract

The implementation of a dual-fuel combustion strategy has recently been explored as a means to improve the thermal efficiencies of internal combustion engines while simultaneously reducing their emissions. Dual-fuel combustion is utilized in compression ignition (CI) engines to promote the use of more readily available gaseous fuels or more efficient, advanced combustion modes. Implementing dual-fuel injection technologies on these engines also allows (1) for improved control of the combustion timing by varying the proportion of two simultaneously injected fuels, and (2) for the use of more advanced combustion modes at high load since the two injected fuels ignite in succession reducing the high peak pressures that generally act as a limiting factor. In spark-ignited (SI) engines, the implementation of a dual-fuel combustion strategy serves as an alternative approach to avoid engine knock. The dual-fuel SI engine relies on the simultaneous injection of a low knock resistance and high knock resistance fuel to dynamically adjust the fuel mixture's resistance to knock as required. The dual-fuel SI engine thereby successfully suppresses knock without compromising the engine efficiency. This chapter discusses the technological advancements associated to dual-fuel combustion and the respective gains in fuel efficiency and emissions reductions that have been achieved.

Keywords: dual-fuel, RCCI, alternative fuels, natural gas, ethanol, knock suppression

1. Introduction

Energy demands in the transportation sector are increasing due to a growing population and simultaneously economic policies are aiming to improve efficiency and reduce hazardous pollutant emissions including nitrogen oxides (NOx), unburned hydrocarbons (UHC) and particulate matter (PM). This has led to a great deal of interest in vehicle electrification as well as cleaner and more efficient engines. While vehicle electrification and hybridization has been growing, the cost and energy density limitations of batteries still pose challenges. As such, it is predicted that internal combustion engines will still power 60% of light-duty vehicles in 2050 [1] and the heavy-duty market will likely be mainly powered by engines for the foreseeable future.

In order to abide by the stringent emissions regulations and deliver power efficiently, there is a need for clean, high efficiency engines. A variety of strategies have been investigated in order to improve the efficiency of today's engines. These include technologies such as variable valve timing that aim to reduce pumping losses associated with the gas exchanges process and variable geometry turbochargers that seek to harness exhaust energy to improve the power density of engines. In addition, more advanced fuel injection systems have also been implemented in order to inject fuel at higher pressures and thereby promote fuel and air mixing. Improved mixing will increase the combustion efficiency and also reduce emissions of particulate matter. More complex fuel injection systems can also be used in order to develop dual-fuel combustion strategies.

Dual-fuel combustion strategies have been demonstrated to be advantageous on both spark-ignited (SI) and compression-ignited (CI) engines. On SI engines, dual-fuel technologies can be leveraged to combat knock. Knock typically occurs in high temperature and high pressure in-cylinder conditions at which the fuel-air mixture will auto-ignite creating pressure shock waves in the cylinder. Knock can significantly damage the engine and is most prevalent at high loads where the efficiency reaches its peak. As such, high efficiency engine performance with gasoline fuel is often limited by knock. In high load conditions, the engine combustion phasing is often delayed to a suboptimal timing in order to avoid knock. While this allows harmful premature combustion to be avoided, it also leads to reductions in efficiency.

Alternatively, knock can also be prevented by using a fuel with a higher octane number (typically described by the research octane number (RON), motor octane number (MON) or anti-knock index (AKI)). Fuels with a high octane rating will be able to operate at the optimal combustion phasing even at high loads, but are more expensive. If high octane fuels are used in dual-fuel engines, they can enable a technique known as "octane-on-demand". Octane-on-demand strategies are often implemented on engines with dual-fuel capabilities by using both a low RON fuel and a high RON fuel simultaneously [2–5]. With dual-fuel capabilities, the fuel mixture's knock resistance can be changed in real time to avoid knock while maintaining optimal combustion phasing. Such methods also allow fuel cost to be minimized since a less expensive, low RON fuel can be used in the lower operating conditions and the high RON fuel can be used only in knock-prone conditions.

On CI engines, dual-fuel injection methods have historically been used for retrofitting old diesel engines with a cheaper fuel. In addition to the utilization of an alternative power source, the implementation also enabled reductions in PM emissions. More recently, dual-fuel injection methods have been used to promote the utilization of less reactive fuels and facilitate more advanced combustion strategies. Some dual-fuel combustion modes have shown significant promise and operate with high efficiency and low pollutant output. This is often achieved over a wide operating range by simultaneously utilizing two fuels with differing reactivities to promote premixing of the fuel or create stratification of the reactivity of the in-cylinder mixture [6, 7].

While these dual-fuel combustion modes show promise, they are not currently utilized in many production vehicles, due to a variety of challenges including difficulties with controlling combustion phasing and combustion stability with the more complex combustion strategy as well as consumer acceptance and infrastructure limitations. Currently, most of these dual-fuel combustion strategies are studied in closely monitored laboratory environments on single cylinder engines. Once removed from the laboratory and implemented on multi-cylinder engines, combustion variations and phasing challenges begin to dominate [8–10]. One such challenge is the occurrence of more significant cylinder-to-cylinder variations that can lead to inconsistent power production and potentially damaging engine conditions. In addition, on CI engines, many dual-fuel combustion strategies leverage a more premixed combustion and as such, the timing of the combustion event is controlled by the chemical kinetics. This makes it more challenging to properly time the combustion event. More advanced control methodologies are required to reduce these combustion variations and ensure an optimal combustion phasing.

Dual-fuel engines have the potential to be highly efficient and clean, but their usage may also be limited by consumer acceptance and infrastructure challenges. Users will have to fill two fuel tanks and will need access to the needed fuels in a broad enough region. This chapter will discuss the technological developments that led to today's dual-fuel engines, and the advancements that have been made on dual-fuel CI and SI engines.

2. Technology overview

The concept of the dual-fuel engine has been around almost as long as the Gasoline (Otto) and Diesel engine. Following the development of Nikolaus Otto's spark-ignited engine, the desire to improve the thermal efficiency by increasing the engine compression ratio led to the development of Rudolf Diesel's compression-ignited engine. Subsequently, interest in better controlling the ignition and regulating the combustion led Rudolf Diesel, himself, to propose a dual-fuel combustion strategy and patent his invention in 1901 [11]. Today, the idea has been leveraged to promote the use of gaseous fuels such as natural gas in diesel engines and for the development of advanced combustion strategies that take advantage of the ability to dynamically optimize the properties of the fuel mixture (by controlling the proportion between the injected fuels) based on the operating conditions. Such implementations of the dual-fuel combustion strategy promise significant gains in fuel efficiency as well as reductions in toxic emissions. Nevertheless, most of the benefits associated with dual-fuel combustion have been primarily explored in academic and research institutions under strictly regulated conditions; the technology currently still faces significant challenges and limited acceptance, which restricts its market penetration.

This section aims to provide an overview of the development of the dual-fuel engines by specifically reviewing the history behind the technology and discussing examples of current and past dual-fuel engines in production. The subsequent sections will discuss ongoing research on dual-fuel engines and its expected role in the near and far future.

2.1 Brief history

In a patent application filed on April 6, 1898, Rudolf Diesel proposes that "if a given mixture is compressed to a degree below its igniting-point, but higher than the igniting-point of a second or auxiliary combustible, then injecting this latter into the first compressed mixture will induce immediate ignition of the secondary fuel and gradual combustion of the first mixture, the combustion after ignition depending on the injection of the igniting or secondary combustible" [11]. This patent entitled *Method of Igniting and Regulating Combustion for Internal Combustion Engines* was accepted in 1901 and marks one of the initial efforts to introduce and successfully ignite a less reactive gaseous fuel in a 4-stroke internal combustion engine using a second fuel. Similarly, today, the ability to ignite a premixed charge (ex: air and a low reactivity fuel such as natural gas) with a secondary high reactivity fuel (such as Diesel) or interchangeably solely operating on the high reactivity fuel is one of the important characterization of a dual-fuel combustion strategy.

For several years, the dual-fuel engine was not used commercially due to its mechanical complexity and rough running caused by auto-ignition and knocking. The first commercial dual-fuel engine was only produced in 1939 by the National Gas and Oil Engine Co. in Great Britain. The engine, fueled by town gas or other types of gaseous fuels, was relatively simple to operate and was mainly employed in some areas where cheap stationary power production was required [12]. During the Second World War, the shortage of liquid fuels attracted further interest in dual-fuel engines from scientists in Great Britain, Germany and Italy. Some diesel engine vehicles were successfully converted to dual-fuel and the possible application of dual-fuel engines in civil and military areas were also explored. Different kinds of gaseous fuels, such as coal gas, sewage gas or methane, were employed in conventional diesel engines during this time [13]. After the Second World War, due to economic and environmental reasons, dual-fuel engines have been further developed and employed in a very wide range of applications from stationary power production to road and marine transport, including long and short haul trucks and busses [12].

In 1949, Crooks, an Engineer at The Cooper-Bessemer Corporation—one of the main engine manufacturers during World War II, presented experimental work with a dual-fuel engine that claimed to have led to the most efficient engine known with a thermal efficiency of 40% at full load. He further highlights that the dual-fuel engine has led to "an extremely economical source of power having an extremely low maintenance cost" [14]. The potential of utilizing relatively cheap gaseous fuel resources and simultaneously benefitting from high thermal efficiencies have promoted the conversion of a conventional compression ignition engine to dual fuel operation. Nevertheless, important limitations still persist: (1) at high loads, the power output and efficiency was limited by the onset of autoignition and knock with most common gaseous fuels, (2) the combustion process in dual-fuel engine is highly sensitive to the type, composition, and concentration of the gaseous fuel being used, and (3) at light load operation, the dual-fuel engine exhibits a greater degree of cyclic variations in performance parameters such as peak cylinder pressure, torque, and ignition delay [13].

A great deal of research is still being undertaken to understand and overcome the challenges associated with the operation of dual-fuel engines. A promising endeavor consists of successfully harnessing the benefits of the dual-fuel engine in the automotive industry.

2.2 Dual-fuel in the modern automotive industry

In a book chapter entitled 'The Dual-fuel Engine' published in 1987, Ghazi A. Karim who had previously conducted several studies [15–20] on the topic of dual-fuel engines suggests that although dual-fuel engine has been employed in a wide range of stationary applications for power production, co-generation, compression of gases and pumping duties; the implementation in mobile applications "remain a field of urgent long term research that can have the potential for opening widely the market for the dual-fuel engine and the increased exploitation of gaseous fuel resources, particularly in the transport sector" [21].

Indeed, the implementation of dual-fuel technology has been more favorable in stationary and heavy-duty applications as opposed to mobile and light-duty applications. Yet, the opportune long-term research proposed by Karim for the transportation sector is still on-going. More recently, efforts to diversify the energy resources of the transportation industry have motivated researchers and engine manufacturers alike to investigate opportunities to leverage the dual-fuel combustion strategy. Furthermore, government imposed regulations on engine-out emissions and fuel efficiency targets have propelled the search for innovative engine technologies including novel implementations of the dual-fuel concept.

In more recent years, a research group at the University of Wisconsin-Madison proposed the implementation of a dual-fuel combustion strategy to reduce Nitrogen Oxide (NOx) and Particulate Matter (PM) emissions [6, 7, 10, 22]. The combustion strategy called Reactivity Controlled Compression Ignition (RCCI) promises significant pollutant reductions as well as impressive fuel efficiency gains. RCCI uses in-cylinder fuel blending with at least two fuels of different reactivity and multiple injections to control in-cylinder fuel reactivity to optimize combustion phasing, duration and magnitude. The process involves introduction of a low reactivity fuel into the cylinder to create a well-mixed charge of low reactivity fuel, air and recirculated exhaust gases. The high reactivity fuel is injected before ignition of the premixed fuel occurs using single or multiple injections directly into the combustion chamber [22].

Kokjohn et al. [6] compared the performance of a conventional diesel combustion and a dual-fuel RCCI combustion. Their study showed the implementation of a dual fuel combustion strategy yielded a reduction in NOx by three orders of magnitude, a reduction in soot by a factor of six, and an increase in gross indicated efficiency of 16.4%. Splitter et al. [7] demonstrated on a dual-fuel RCCI engine that optimizing in-cylinder fuel stratification with two fuels of large reactivity differences achieved gross indicated thermal efficiencies near 60%. Furthermore, they showed through simulations studies that a dual-fuel combustion strategy rejected less heat, and that ~94% of the maximum cycle efficiency could be achieved while simultaneously obtaining ultra-low NOx and PM emissions.

Similar motivations to boost the thermal efficiency of engines have led to the implementation of a dual-fuel strategy in light duty-spark ignited engines as well. Initially proposed as an engine concept in 2005 by Cohn et al. [3], the dual-fuel spark-ignited engine featured two fuel injections systems—one for conventional gasoline and another for ethanol. The engine would promote the utilization of alternative fuels such as ethanol reducing the dependence on fossil fuels, and it was an alternative knock suppressing strategy which allows for higher load and higher efficiency operations. A high octane rating fuel such as ethanol is used in conjunction with the conventional fuel, gasoline, to dynamically adjust the fuel mixture's resistance to auto-ignition based on the operating conditions.

The studies by Cohn et al. [3] suggested dual-fuel combustion could potentially increase an SI engine's drive cycle efficiency by approximately 30%. Similar studies by Daniel et al. [4] demonstrated that dual-injection showed benefits to the indicated efficiency and emissions at almost all loads compared to a single fuel gasoline direct injection (GDI) strategy. Furthermore, Chang et al. [77] showed a maximum 30% Well-to-Wheels (W-t-W) CO_2 equivalent reduction can be achieved by utilizing a dual-fuel injection system. Numerous studies, such as [23–25], continue to explore the benefits that can be achieved through the introduction of dual-fuel combustion in the modern automotive engines.

In the next sections, the application and benefits of dual-fuel combustion are separately discussed for compression-ignition and spark-ignited engines followed by concluding remarks.

3. Dual-fuel compression-ignition engines

Diesel or compression-ignition engines dominate the medium and heavy-duty markets due to their higher efficiency and high torque production capabilities. Such engines require a more reactive fuel that will auto-ignite at high pressures and temperatures. This limits the fuels that can be leveraged on CI engines. Dual-fuel engines provide a way to use less reactive fuels since they can leverage a second more reactive fuel to produce ignition. In addition, dual-fuel concepts have also been investigated as a way to reduce engine emissions. Conventional diesel combustion is diffusion controlled and is typically accompanied by high nitrogen oxide

(NO$_x$) and particulate matter (PM) emissions [26]. Nitrogen oxide emissions result from high in-cylinder temperature conditions which promote the combination of nitrogen (carried in with the fresh air) with excess oxygen [27]. Meanwhile, particulate matter or soot is produced in fuel rich regions when hydrocarbon species agglomerate [27, 28]. As such, high local equivalence ratios can lead to soot formation and high local temperatures can lead to NOx formation as shown in **Figure 1**. In order to avoid these problematic regions, many dual-fuel, heavy-duty CI engines attempt to operate in conditions which promote premixing of the fuel and air and/ or achieve in-cylinder stratification in order to reach high efficiencies and low emissions. By enabling a more premixed combustion, rich regions where PM would be produced can be nearly eliminated and shorter combustion durations are achieved which reduces local temperatures and thereby, NO$_x$ emissions [6, 7, 29–33].

As such, dual-fuel engines have been pursued in the heavy-duty market for two main reasons:

1. As a way to leverage more readily available but less reactive fuels as the primary power source and use a high reactivity fuel to initiate combustion.

2. As a way to introduce fuels of varying reactivities and create a more complex combustion mode that can be more efficient and produce less NO$_x$ and PM.

3.1 Conventional dual-fuel compression-ignition engines

As the world seeks to become less reliant on conventional diesel and gasoline, there has been increasing interest in using fuels such as natural gas in engines. Some of these fuels are less reactive than conventional diesel fuel and therefore, are more challenging to use on compression-ignition engines where auto-ignition of the fuel is needed. Dual-fuel systems are one way to leverage less reactive fuels on heavy-duty engines [34–38]. One such fuel is natural gas and it will be focused on here as an example of the benefits and challenges of this type of engine operation.

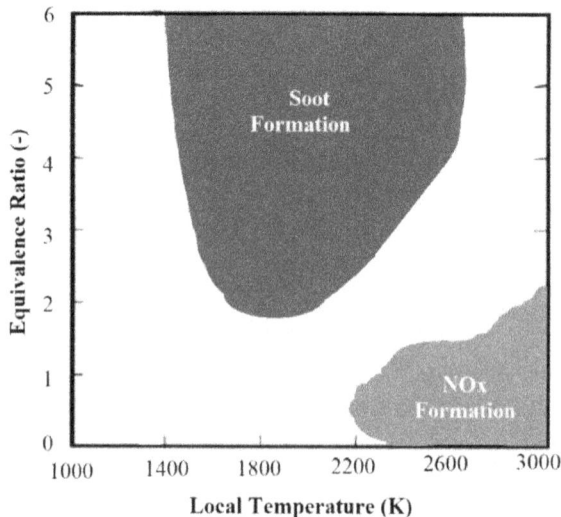

Figure 1.
Emissions with respect to local temperature versus local equivalence ratio.

Natural gas is more difficult to ignite than conventional engine fuels so it is more easily integrated into spark-ignited engines. On heavy-duty engines, natural gas needs an ignition source so it is typically port-injected and diesel is direct injected and serves as a pilot. Fuels that are port-injected become premixed with the air and typically exhibit a rapid combustion event that is dominated by the chemical kinetics of the combustion reaction, but fuels that are direct injected and have to mix with the air tend to have a longer combustion event that is dominated by the time taken for the air and fuel to mix adequately. Since dual-fuel engines have a fuel that is port-injected and one that is direct-injected, they often exhibit a two-stage combustion process. The portion of combustion that occurs in a premixed vs. a diffusion mode will be strongly dependent on the amount of each fuel that is used [39]. While this makes the combustion process more complicated, dual-fuel injection can provide stable combustion of a less reactive fuel like natural gas in CI engines. However, fuel economy reductions around 10% have been observed when operating in this type of mode [34].

Not only is fuel economy or efficiency impacted, but emissions are also altered with dual-fuel combustion. In natural gas-diesel dual-fuel engines, up to 60% reductions in NOx and PM have been observed [34]. However, these emissions are dependent on the fuels used as well as the amounts of each fuels used. For example, particulate matter emissions and the particle size distribution of the particulates have been shown to strongly depend on the properties of the direct-injected fuel and level of natural gas substitution. Direct injected fuels with lower densities and viscosity and higher volatility produce lower amounts of particulates [40]. However, higher natural gas substitution rates can increase soot levels since they decrease the local oxygen availability [41].

As with many natural gas engines, higher CO and UHC emissions are typically encountered. Various natural gas substitution rates have been explored in [42] and showed that only lower amounts of natural gas could be used at low load conditions due to emissions constraints, but higher fractions of natural gas could be used at high loads. Direct injection of both fuels [43], higher fuel injection pressures, and adapted engine control units [44, 45] have been implemented to avoid these emissions constraints. After treatment systems including diesel oxidation catalysts [35] as well as diesel particulate filters and urea-selective catalytic reduction systems [46] have also been introduced on dual-fuel engines to reduce emissions. However, to enable efficient use of high amounts of natural gas, more advanced combustion methods and optimization methods are likely needed [47, 48].

A majority of conventional dual-fuel engine studies have focused on natural gas, but this approach of using diesel as a pilot fuel can also be leveraged with a variety of fuels that are not reactive enough to be used as the sole fuel on a compression-ignition engine. Dual-fuel concepts have also been explored with fuel combinations including on methanol and diesel [49], biogas and biodiesel and biogas and diesel [50].

3.2 Advanced dual-fuel compression-ignition engines

In order to push engines to higher efficiencies, there has been a great deal of exploration into more complex combustion modes. Many of these advanced combustion strategies attempted to premix the fuel and air in order to achieve a more efficient and clean combustion, but were only able to be leveraged in lower torque ranges [51, 52]. One strategy for expanding the operating region of these more advanced techniques is to simultaneously utilize two fuels with differing reactivities in order to further increase the combustion delay period and promote premixing in higher operating regions [53]. This strategy is known as reactivity-controlled compression ignition (RCCI). In RCCI, a fuel with low reactivity such as gasoline is injected separately from a high reactivity diesel-type fuel. The quantities of each respective fuel can be modified so that the combustion event can be delayed to provide adequate mixing time and the desired shape of the combustion event can be achieved. Recent work in RCCI has shown that fuel properties that differ from those of conventional fuels can be leveraged to shape the combustion process and increase engine efficiency from 45% to near 60% [6, 7] in this mode. While the efficiency benefits can be significant, high CO and UHC emissions as well as high pressure rise rates can still limit the use of RCCI.

3.2.1 Reactivity controlled compression ignition

RCCI-type combustion was originally studied at the University of Wisconsin-Madison using gasoline as the port-injected low reactivity fuel and diesel as the direct-injected, high reactivity fuel [7]. By leveraging two fuels with varying properties stratification of the in-cylinder mixture reactivity could be achieved leading to longer ignition delays and increased time for premixing. Diesel fuels with lower reactivities were shown to be advantageous in these operating conditions as they increase the local reactivity gradient [54, 55]. In such modes, the more reactive fuel components are consumed at a faster rate and the slower burning competent make up a larger portion of the UHC emissions [56].

The use of alternative fuels such as ethanol and natural gas in such RCCI-type operation conditions has also shown promising results and appears to better take advantage of these alternatives. Research by Navistar, Argonne National Laboratory and Wisconsin Engine Research Consultants found that using E85 as the low reactivity fuel could allow higher loads and efficiencies to be achieved with RCCI. While more traditional gasoline and diesel dual-fuel operation achieved a BMEP of 11.6 bar and brake thermal efficiency (BTE) of 43.6%, using E85 with diesel allowed operation to be extended to 19 bar BMEP with a BTE of 45.1% [57]. Later studies led by RWTH Aachen University and FEV GmbH showed that when using diesel and ethanol, higher ethanol quantities could be leveraged in lower load conditions and would provide a more stable combustion and lower UHC emissions. However, as the load was increased higher amounts of diesel were required in order to keep the cylinder pressure rise rate to an acceptable level [58].

Some of these detrimental impacts on CO and UHC emissions are able to be counteracted by more complex fuel injection strategies [59]. For example, more recent work has explored the use of ethanol port-injection with a multi-pulse direct-injection of diesel. A double pilot injection was able to reduce the UHC and CO emissions in RCCI-type conditions [60]. Other methods such as leveraging higher injection pressures have also been shown to be able to increase efficiency and provide further decreases in emissions of NOx, CO, UHC, and PM [61].

3.2.2 Challenges with reactivity controlled compression ignition

While RCCI methods are promising, these modes suffer from several technical challenges. First, cycle-to-cycle and cylinder-to-cylinder variations can be more dramatic than in conventional diesel combustion [62, 63]. Since fuel and air mixing are critical and high amounts of recirculated exhaust gas are typically leveraged in these modes, small variations in the in-cylinder fuel quantities and gas mixture can lead to significant variations in the combustion process. Combating such variations is likely to require more complex control strategies and additional engine sensors [62].

Second, control of the combustion phasing of these modes is challenging since the combustion process is controlled by chemical kinetics and not directly triggered by an injection event. Control techniques must try to maintain an optimal combustion phasing while ensuring that pressure rise rate and combustion variations do not exceed acceptable limits by monitoring the fuel blend ratio [64] and direct-injection timing [65]. Successful use of RCCI may also require switching between traditional diesel combustion at lower load and dual-fuel operation at higher load [65]. Intermediate modes such as "premixed dual-fuel combustion" and "partially premixed compression ignition" may provide clean and efficient intermediate combustion strategies that can be used on their own or in transitions from conventional diesel combustion to RCCI [66].

As discussed previously, UHC and CO emissions are often higher in RCCI modes. This is believed to be because local equivalence ratios can drop below the flammability limit of natural gas and lead to unburned hydrocarbon emissions [67], but may necessitate the development of new after treatment systems for these engines. Consumer acceptance is also a concern with dual-fuel engines. Since users may not want to fill two fuel tanks, Splitter et al. explored a method of enabling RCCI by using gasoline and the cetane improver di-tert-dutyl peroide (DTBP) [22]. This study leveraged port injected gasoline as the low reactivity fuel, but used gasoline mixed with varying amounts of DTBP as the high reactivity fuel. A peak gross ITE of 57% was achieved and emissions were similar to standard dual-fuel RCCI levels.

4. Dual-fuel spark-ignited engines

The implementation of dual-fuel combustion strategies on medium and heavy-duty engines have primarily been discussed in the preceding section. These utilizations of dual-fuel combustion strategies are generally restricted to compression-ignition engines where Diesel is the conventional fuel. Nevertheless, in recent years, light-duty spark ignited engines have also been configured to feature a dual-fuel combustion system. This section will discuss the implementation and utilization of dual-fuel combustion on light-duty SI engines.

In SI engines, a dual-fuel combustion strategy is also leveraged to promote the utilization of alternative fuels such as ethanol and methanol. There are increasingly numerous government imposed legislations promoting the use of biofuels in transportation [68]. Current legislation requires EU member states to conform to a 10% minimum target on the use of alternative fuels (biofuels or other renewable fuels) in transportation by 2020 [69] . In the US, tax incentives have been used to promote the use of ethanol in gasoline [70], in order to replicate the success seen in Brazil [71]. The quest to benefit from incentives or conform to legislations has led engine manufacturers to explore the implementation of dual-fuel combustion on spark-ignited engines with the utilization of biofuels and other renewable fuels.

Additionally, one of the primary motivations for the implementation of dual-fuel combustion in SI engines has also been for the development of better engine knock control techniques. Engine knock, the inadvertent auto-ignition of the fuel in localized high pressure and temperature regions inside the cylinder [72, 73], can result in significant engine damage and marks one of the main obstacles in SI engines. The conventional approach to avoiding knock in spark-ignited (SI) engines consists of delaying the combustion phasing by retarding the spark timing [74]. A combustion event occurring later in the combustion stroke (further away from top dead center) has a lower tendency to knock since the combustion pressure and temperature are lower. However, delaying the combustion phasing also reduces the fuel efficiency since less work can be extracted by the late combustion [75].

A dual-fuel combustion strategy provides SI engines with an alternative way to avoid knock without sacrificing fuel efficiency. The tendency of the fuel to auto-ignite is not only dependent on the in-cylinder conditions, but also on the knock resistance (octane rating) of the fuel. As such, increasing the fuel's octane rating helps avoid knock without compromising fuel efficiency. Engines with dual-fuel capabilities can use a low RON fuel and a high RON fuel simultaneously to optimize the fuel mixture's knock resistance by controlling the proportion of each injected fuel. Many studies have explored the implementation of a dual-fuel strategy to suppress knock [2–5, 72–75].

The studies by Cohn et al. [3] and Bromberg et al. [76] at Massachusetts Institute of Technology proposed an ethanol boosted engine concept, which provides suppression of engine knock at high pressure through the use of direct ethanol injection. Their studies conclude that the implementation of the secondary fuel injection system could allow engine operation at much higher levels of turbocharging and could potentially increase the drive cycle efficiency by approximately 30%. Daniel et al. [4] implemented a dual-fuel strategy for knock mitigation on a single cylinder SI research engine. The study shows that dual-injection strategy (using either ethanol or methanol as the high RON fuels) showed benefits to the indicated efficiency and emissions (HC, CO, CO_2) at almost all loads compared to a single fuel gasoline direct injection (GDI) strategy. Furthermore, Chang et al. [77] conducted a Well-to-Wheels (W-t-W) greenhouse gas emissions assessment to estimate the overall emissions benefits of a knock mitigating dual-fuel system. Their study showed a maximum 30% W-t-W CO_2 equivalent reduction can be achieved by utilizing a dual-fuel injection system.

A dual-fuel SI configuration provides three main benefits: (1) engine can be further downsized and operated in high pressure conditions (2) the fuel knock resistance can be adjusted based on operating point while maintaining an optimal combustion phasing (maximizing engine efficiency), and (3) operating points with low knock propensity can be operated with a low octane fuel, eliminating the waste of RON, which generally translates into cheaper fuel cost and lower CO_2 emissions [5, 77]. A team at Saudi Aramco, in collaboration with IFP Energies nouvelles, demonstrated these benefits on a production passenger vehicle [23, 78, 79]. The dual-fuel technology is identified as "an opportunity to improve fuel efficiency by using the octane only when you need it." The researchers outline that the technology will improve fuel efficiency while reducing overall energy requirements to manufacture gasoline fuels in the future [79]. The researchers at Aramco Fuel Research Center (AFRC) identify the development of the production car as only the start, and outline the near-term objective of going from a vehicle with two tanks with two different fuels to one that only uses only one fuel and process it with an on-board fuel upgrading system.

Similar to the efforts of Saudi Aramco, there is currently a growing interest to harness the benefits of a dual-fuel combustion strategy on conventional SI engines. A study at Massachusetts Institute of Technology by Jo et al. [24] investigates the use of dual-fuel for a passenger vehicle and a medium-duty truck. Their simulation studies, coupled with experimental testing, conclude that significant gains in engine brake efficiencies can be achieved: 30% for the Urban Dynamometer Driving Schedule (UDDS) cycle and 15% for the US06 cycle. Studies by Marchitto et al. at Istituto Motori, CNR [25] demonstrate the port injection of ethanol as a secondary fuel showed a significant increase in thermal efficiency (~10%) and significant reduction in particle number emissions as well as particulate mass (60–80%).

Dual-fuel combustion provides a promising path to boost the thermal efficiency of spark-ignited engines. The opportunity to leverage alternative fuels such as ethanol and methanol, as well as the ability to suppress knock without compromising thermal efficiency, has garnered interest in the technology. Nevertheless, the challenges associated with multiple tanks and multiple fuels require researchers to seek paths that will make the technology more accessible to everyday consumers.

5. Conclusion

Since the inception of the dual-fuel combustion strategy as a tool to better control combustion, its application has been most vital in stationary and heavy-duty applications. The integration of the dual-fuel combustion in the transportation industry promises great benefits both in terms of fuel efficiency improvement as well as toxic emissions reduction. Significant efforts are undertaken to implement this technology in the automotive industry for heavy-duty, as well as medium and light-duty engines. The on-going research on dual-fuel combustion promises encouraging paths that will allow the utilization of more readily available gaseous fuels and renewable fuels. The observed benefits on both compression-ignition and spark-ignited engines warrants further investments, research, and efforts to better exploit these gains on a larger scale.

In compression ignition (CI) engines, dual-fuel combustion presents an effective approach to control combustion timing and extend engine load limitations. This is achieved by injecting both a high reactivity and low reactivity fuel, adjusting the concentration of one relative to the other, and thereby optimizing the fuel mixture's reactivity (on a cycle-by-cycle basis) for different operating conditions. In spark ignited (SI) engines, the optimization of the fuel's octane rating presents an alternative approach to avoid abnormal combustion (engine knock due to auto-ignition). The conventional SI engine relies on delayed spark timings to avoid knock, which results in degraded efficiency and higher emissions. With dual-fuel combustion techniques, two fuels with high and low octane ratings can be used to adjust the fuel mixture's octane as needed to avoid knock without sacrificing the engine's efficiency.

While dual-fuel operation has many potential benefits, the implementation of these strategies on CI and SI engines also involve significant challenges. These challenges are exacerbated when dual-fuel combustion is implemented in conjunction with other advanced combustion strategies including varying valve timing, and high EGR circulation. The underlying challenges include increased combustion variations and difficulties in properly adjusting fuel mixture for effective control of combustion timing (in CI engines) and effective knock control (in SI engines). While the technology has successfully been used in stationary applications, the implementation of dual-fuel strategy on mobile applications, specifically in the transportation sector, still faces limiting challenges. In addition to the technical challenges associated with the dual-fuel engine, a primary concern for its integration in the automotive industry consists of the social resistance to the requirement of having and filling two fuel tanks. In order for the technology to successfully penetrate the automotive market, the benefit in the terms of fuel efficiency improvement and toxic emissions reduction need to clearly outweigh the technical and social challenges.

Author details

Mateos Kassa[1*] and Carrie Hall[2]

1 IFP Energies Nouvelles, Rueil-Malmaison, France

2 Illinois Institute of Technology, Chicago, IL, USA

*Address all correspondence to: mateos.kassa@ifpen.fr

IntechOpen

References

[1] Heywood J. Not dead yet: The resilient ICE looks to 2050. Automotive Engineering, SAE Internation. April 2018

[2] Baranski J, Anderson E, Grinstead K, Hoke J, Litke P. Control of fuel octane for knock mitigation on a dual-fuel spark-ignition engine. SAE Technical Paper 2013-01-0320. 2013

[3] Cohn DR, Bromberg L, Heywood JB. Direct Injection Ethanol Boosted Gasoline Engines: Biofuel Leveraging for Cost Effective Reduction of Oil Dependence and CO_2 Emissions. Cambridge, MA: Massachusetts Institute of Technology; 2005

[4] Daniel R, Wang C, Xu H, Tia G, Richardson D. Dual-injection as a knock mitigation strategy using pure ethanol and methanol. SAE International Journal of Fuels and Lubricants. 2012;5:772-784

[5] Viollet Y, Abdullah M, Alhajhouje A, Chang J. Characterization of high efficiency octane-on-demand fuels requirement in a modern spark ignition engine with dual injection system. SAE Technical Paper 2015-01-1265. 2015

[6] Kokjohn S, Hanson R, Splitter D, Reitz R. Fuel reactivity controlled compression ignition (RCCI): A pathway to controlled high-efficiency clean combustion. International Journal of Engine Research. 2011;12(3):209-226

[7] Splitter D, Wissink M, Del Vescovo D, Reitz R. RCCI engine operation towards 60% thermal efficiency. SAE 2013-01-0279. 2013

[8] Beatrice C, Avolio G, Beroli C, Del Giacomo N, et al. Critical aspects on the control in the low temperature combustion systems for high performance DI diesel engines. Oil & Gas Science Technology. 2007;62(4):471-482

[9] Bittle J, Zheng J, Xue X, Song H, Jacobs T. Cylinder-to-cylinder variation sources in diesel low temperature combustion and the influence they have on emissions. International Journal of Engine Research. 2014;15(1):112-122

[10] Klos DT, Kokjohn SL. Investigation of the effect of injection and control strategies on combustion instability in reactivity controlled compression ignition engines. Journal of Engineering Gas Turbines Power. 2015;138(1)

[11] Diesel R. Method of igniting and regulating combustion for internal combustion engines. U.S. Patent 673,160, April 1901

[12] Sahoo BB, Sahoo N, Saha UK. Effect of engine parameters and type of gaseous fuel on the performance of dual-fuel gas diesel engines—A critical review. Renewable and Sustainable Energy Reviews. 2009;13:1151-1184

[13] Liu Z. An Examination of the Combustion Characteristics of Compression Ignition Engines Fuelled with Gaseous Fuels. Ph.D, University of Calgary; 1995

[14] Crooks WR. The dual-fuel engine and its application to sewage treatment plants. Sewage Works Journal. 1949;21:957-961

[15] Karim GA, Klat SR, Moore NPW. Knock in dual-fuel engines. Proceedings of the Institution of Mechanical Engineers. 1966;181:453-466

[16] Karim GA, Khan MO. Examination of effective rates of combustion heat release in a dual-fuel engine. Journal of Mechanical Engineering Science. 1968;10:13-23

[17] Karim GA. The ignition of a premixed fuel and air charge by pilot

fuel spray injection with reference to dual-fuel combustion. SAE Technical Paper Series; 1968. DOI: 10.4271/680768

[18] Karim GA. A review of combustion processes in the dual fuel engine—The gas diesel engine. Progress in Energy and Combustion Science. 1980;**6**:277-285

[19] Karim GA, Burn KS. The Combustion of Gaseous Fuels in a Dual Fuel Engine of the Compression Ignition Type with Particular Reference to Cold Intake Temperature Conditions. SAE Technical Paper Series; 1980. DOI: 10.4271/800263

[20] Karim GA. The Dual Fuel Engine of the Compression Ignition Type - Prospects, Problems and Solutions - A Review. SAE Technical Paper Series; 1983. DOI: 10.4271/831073

[21] Karim GA. The dual fuel engine. In: Evans RL, editor. Automotive Engine Alternatives. Boston, MA: Springer US; 1987. pp. 83-104

[22] Splitter D, Reitz R, Hanson R. High efficiency, low emissions RCCI combustion by use of a fuel additive. SAE International Journal of Fuels and Lubricants. 2010;**3**(2):742-756

[23] Morganti K, Viollet Y, Head R, Kalghatgi G, Al-Abdullah M, Alzubail A. Maximizing the benefits of high octane fuels in spark-ignition engines. Fuel. 2017;**207**:470-487

[24] Jo YS, Bromberg L, Heywood J. Optimal use of ethanol in dual fuel applications: Effects of engine downsizing, spark retard, and compression ratio on fuel economy. SAE International Journal of Engines. 2016;**9**:1087-1101

[25] Marchitto L, Tornatore C, Costagliola MA, Valentino G. Impact of Ethanol-Gasoline Port Injected On Performance and Exhaust Emissions of A Turbocharged SI Engine.

SAE Technical Paper Series; 2018. DOI: 10.4271/2018-01-0914

[26] Heywood J. Internal Combustion Engine Fundamentals. New York, NY, USA: McGraw-Hill; 1998

[27] Turns S. An Introduction to Combustion: Concepts and Applications. New Dehli, India: McGraw Hill; 2013

[28] Mansurov ZA. Soot formation in combustion processes (review). Combustion, Explosion and Shock Waves. 2005;**41**(6):727-744

[29] Akihama K, Takatori Y, Inagaki K, Sasaki S, Dean AM. Mechanism of the smokeless rich diesel combustion by reducing temperature. SAE International Paper 2001-01-0655. 2001

[30] Neely GD, Sasaki S, Huang Y, Leet JA, Stewart DW. New diesel emission control strategy to meet US tier 2 emissions regulations. SAE International Paper 2005-01-1091. 2005

[31] Caton JA. The thermodynamic characteristics of high efficiency, internal-combustion engines. Energy Conversion and Management. 2012;**58**:84-93

[32] Ma J, Lu X, Ji L, Huang Z. An experimental study of HCCI-DI combustion and emissions in a diesel engine with dual fuel. International Journal of Thermal Sciences. 2008;**47**:1235-1242

[33] Leermakers C, Luijten C, Somers L, Kalghatgi G. Experimental study of fuel composition impact on PCCI combustion in a heavy-duty diesel engine. SAE 2011-01-1351. 2011

[34] Addy JM, Bining A, Norton P, Peterson E, Campbell K, Bevillaqua O. Demonstration of Caterpillar C10 dual fuel natural gas engines in

commuter buses. SAE Technical Paper 2000-01-1386. 2000

[35] Mittal M, Donahue R, Winnie P, Gillette A. Combustion and gaseous emissions characteristics of a six-cylinder diesel engine operating within wide range of natural gas substitutions at different operating conditions for generation application. SAE Technical Paper 2014-01-1312. 2014

[36] Taritas I, Kozarac D, Sjeric M, Aznar MS, Vuilleumier D, Tatschl R. Development and validation of a quasi-dimensional dual fuel (diesel-natural gas) combustion model. SAE International Journal of Engines. 2017;**10**(2):2017-01-0517

[37] Wurzenberger JC, Katrasnik T. Dual fuel engine simulation—A thermodynamic consistent HiL compatible model. SAE International Journal of Engines. 2014;**7**(1):2014-01-1094

[38] Hountalas RPD. Combustion and exhaust emission characteristics of a dual fuel compression ignition engine operated with pilot diesel fuel and natural gas. Energy Conversion and Management. 2004;**45**(18-19):2971-2987

[39] Ahmad Z, Aryal J, Ranta O, Kaario O, Vuorinen V, Larmi M. An optical characterization of dual-fuel combustion in a heavy-duty diesel engine. SAE Technical Paper 2018-01-0252. 2018

[40] Zhang Y, Ghandhi J, Rothamer D. Effects of fuel chemistry and spray properties on particulate size distributions from dual-fuel combustion strategies. SAE International Journal of Engines. 2017;**10**(4):2017-01-1005

[41] Srna A, Bruneaux G, Rotz BV, Bombach R, Herrmann K, Boulouchos K. Optical investigation of sooting propensity of n-dodecane pilot/lean-premixed methane dual-fuel combustion in a rapid compression-expansion machine. SAE Technical Paper 2018-01-0258. 2018

[42] Garcia P, Tunestal P. Experimental investigation on CNG-diesel combustion modes under highly diluted conditions on a light duty diesel engine with focus on injection strategy. SAE International Journal of Engines. 2015;**8**(5):2015-24-2439

[43] Fasching P, Sprenger F, Eichlseder H. Experimental optimization of a small bore natural gas-diesel dual fuel engine with direct fuel injection. SAE International Journal of Engines. 2016;**9**(2):2016-01-0783

[44] Yang B, Wei X, Zeng K, Lai M-C. The development of an electronic control unit for a high pressure common rail diesel/natural gas dual-fuel engine. SAE Technical Paper 2014-01-1168. 2014

[45] Xu S, Anderson D, Singh A, Hoffman M, Prucka R, Filipi Z. Development of a phenomenological dual-fuel natural gas diesel engine simulation and its use for analysis of transient operations. SAE International Journal of Engines. 2014;**7**(4):2014-01-2546

[46] Besch MC, Israel J, Thiruvengadam A, Kappanna H, Carder D. Emissions characterization from different technology heavy-duty engines retrofitted for CNG/diesel dual-fuel operation. SAE Intenrational Journal of Engines. 2015;**8**(3):2015-01-1085

[47] Mattson JMS, Langness C, Depcik C. An analysis of dual-fuel combustion of diesel with compressed natural gas in a single-cylinder engine. SAE Technical Paper 2018-01-0248. 2018

[48] Kozarac D, Sjeric M, Krajnovic J, Sremec M. The optimization of the dual fuel engine injection parameters by using a newly developed quasi-dimensional cycle simulation

combustion model. SAE Technical Paper 2018-01-0261. 2018

[49] Saccullo M, Benham T, Denbratt I. Dual fuel methanol and diesel direct injection HD single cylinder engine tests. SAE Technical Paper 2018-01-0259. 2018

[50] Yoon SH, Lee CS. Experimental investigation on the combustion and exhaust emission characteristics of biogas-biodiesel dual-fuel combustion in a CI engine. Fuel Processing Technology. 2011;**92**(5):992-1000

[51] Keeler B. Constraints on the operation of a DI diesel [PhD thesis]. The University of Nottingham; 2009

[52] Kulkarni AM, Stricke KC, Blum A, Shaver GM. PCCI control Authority of a Modern Diesel Engine Outfitted with Flexible Intake Valve Actuation. Journal of Dynamic Systems, Measurement, and Control. 2010;**132**(5)

[53] Kokjohn SL, Hanson RM, Splitter DA, Reitz RD. Experiments and modeling of dual-fuel HCCI and PCCI combustion using in-cylinder fuel blending. SAE Internation Journal of Engines. 2010;**2**(2):24-39

[54] Ickes A, Wallner T, Zhang Y, Ojeda WD. Impact of Cetane number on combustion of a gasoline-diesel dual-fuel heavy-duty multi-cylinder engine. SAE International Journal of Engines. 2014;**7**(2):2014-01-1309

[55] Benajes J, Garcia A, Monsalve-Serrano J, Boronat V. Influence of direct-injected fuel properties on performance and emissions from a light-duty diesel engine running under RCCI combustion mode. SAE Technical Paper 2018-01-0250. 2018

[56] Puduppakkam KV, Liang L, Naik CV, Meeks E, Kokjohn SL, Reitz RD. Use of detailed kinetics and advanced chemistry-solution techniques in CFD to investigate dual-fuel engine concepts. SAE Technical Paper No. 2011-01-0895:2011

[57] Zhang Y, Sagalovich I, Ojeda WD, Ickes A, Wallner T, Wickman DD. Development of dual-fuel low temperature combustion strategy in a multi-cylinder heavy-duty compression ignition engine using conventional and alternative fuels. SAE International Journal of Engines. 2013;**6**(3):2013-01-2422

[58] Heuser B, Kremer F, Pischinger S, Rohs H, Holderbaum B, Körfer T. An experimental investigation of dual-fuel combustion in a light duty diesel engine by in-cylinder blending of ethanol and diesel. SAE International Journal of Engines. 2015;**9**(1):2015-01-1801

[59] Nithyanandan K, Hou D, Major G, Lee C-F. Spray visualization and characterization of a dual-fuel injector using diesel and gasoline. SAE International Journal of Fuels and Lubricants. 2014;**7**(1):2014-01-1403

[60] Yu S, Dev S, Yang Z, Leblanc S, Yu X, et al. Early pilot Injection strategies for reactivity control in diesel-ethanol dual fuel combustion. SAE Technical Paper 2018-01-0265. 2018

[61] Mobasheri R, Seddiq M. Effects of diesel injection parameters in a heavy duty iso-butanol/diesel reactivity controlled compression ignition (RCCI) engine. SAE Technical Paper 2018-01-0197. 2018

[62] Kassa M, Hall C, Ickes A, Wallner T. Modeling and control of fuel distribution in a dual fuel internal combustion engine leveraging late intake valve closings. International Journal of Engine Research. 2016;**18**(8):797-809

[63] Dong S, Ou B, Cheng X. Comparisons of the cyclic variability of gasoline/diesel and ethanol/diesel dual-fuel combustion based on a diesel engine. SAE Technical Paper 2017-01-5001. 2017

[64] Han X, Divekar P, Reader G, Zheng M, Tjong J. Active injection control for enabling clean combustion in ethanol-diesel dual-fuel mode. SAE International Journal of Engines. 2015;**8**(2):2015-01-0858

[65] Divekar P, Han X, Tan Q, Asad U, Yanai T, Chen X, et al. Mode switching to improve low load efficiency of an ethanol-diesel dual-fuel engine. SAE Technical Paper 2017-01-0771. 2017

[66] Martin J, Boehman A, Topkar R, Chopra S, Subramaniam U, Chen H. Intermediate combustion modes between conventional diesel and RCCI. SAE Technical Paper 2018-01-0249. 2018

[67] Sprenger F, Fasching P, Kammerstätter S. Experimental investigation of CNG-diesel combustion processes with external and internal mixture formation for passenger Car applications. In: Proceedings of the Conference on the Working Process of the Internal Combustion Engine. Austria: Graz; September 2015. pp. 24-25

[68] Wu X, Daniel R, Tian G, Xu H, Huang Z, Richardson D. Dual-injection: The flexible, bi-fuel concept for spark-ignition engines fuelled with various gasoline and biofuel blends. Applied Energy. 2011;**88**:2305-2314

[69] Union E. Directive 2009/28/EC of the European Parliament and of the Council of 23 April 2009 on the promotion of the use of energy from renewable sources and amending and subsequently repealing Directives 2001/77/EC and 2003/30/EC. Official Journal of the European Union. 2009;**5**:2009

[70] Curtis B. US Ethanol Industry: The Next Inflection Point. 2007 Year in Review. San Francisco, CA: BCurtis Energies & Resource Group; 2008. p. 52

[71] Goldemberg J. The challenge of biofuels. Energy & Environmental Science. 2008;**1**:523-525

[72] Haskell WW, Bame JL. Engine Knock–An End-Gas Explosion. SAE Technical Paper Series; 1965. DOI: 10.4271/650506

[73] Zhen X, Wang Y, Xu S, Zhu Y, Tao C, et al. The engine knock analysis—An overview. Applied Energy. 2012;**92**:628-636

[74] Guzzella L, Onder C. Control of engine systems: Engine knock. In: Introduction to Modeling and Control of Internal Combustion Engine Systems. Berlin: Springer Berlin; 2014. pp. 199-209

[75] Ayala FA, Gerty MD, Heywood JB. Effects of combustion phasing, relative air-fuel ratio, compression ratio, and load on SI engine efficiency. SAE Technical Paper Series; 2006. DOI: 10.4271/2006-01-0229

[76] Bromberg L, Cohn DR, Heywood JB. Calculations of knock suppression in highly turbocharged gasoline/ethanol engines using direct ethanol injection. Massachusetts Institute of Technology; 2006

[77] Chang J, Viollet Y, Alzubail A, Abdul-Manan AFN, Al Arfaj A. Octane-on-Demand as an Enabler for Highly Efficient Spark Ignition Engines and Greenhouse Gas Emissions Improvement. SAE Technical Paper Series, 2015. DOI: 10.4271/2015-01-1264

[78] Saudi Aramco—A Step Forward in Fuel Technology [Online]. 2017. Available from: http://www.saudiaramco.com/en/home/news-media/news/a-step-forward-in-fuel-technology.html [Accessed: June 10, 2018]

[79] Green Car Congress [Online]. 2017. Available from: http://www.greencarcongress.com/2017/08/20170810-aramco.html [Accessed: June 10, 2018]

Chapter 3

Alternative Fuels for Internal Combustion Engines

Mehmet Ilhan Ilhak, Selim Tangoz,
Selahaddin Orhan Akansu and Nafiz Kahraman

Abstract

Researchers have studied on alternative fuels that can be used with gasoline and diesel fuels. Alternative fuels such as hydrogen, acetylene, natural gas, ethanol and biofuels also uses in internal combustion engines. Hydrogen in the gas phase is about 14 times lighter than the air. Moreover, it is the cleanest fuel in the world. On the other hand because of its high ignition limit (4–75%), low ignition energy, needs special design to use as pure hydrogen in internal combustion engines. It is proved that hydrogen improves the combustion, emissions and performance, when is added as 20% to fuels. Natural gas is generally consisting of methane (85–96%) and it can be used in both petrol and diesel engines. Ethanol can be used as pure fuel or mixed with different fuels in internal combustion engines. In this section, the effects of natural gas, hydrogen, natural gas + hydrogen (HCNG), ethanol, ethanol + gasoline, ethanol + hydrogen, acetylene, acetylene + gasoline mixtures on engine performance and emissions have been examined.

Keywords: internal combustion engines, hydrogen, acetylene, natural gas, ethanol

1. Introduction

Oil is the undisputed largest source of energy for internal combustion engines (ICE). However, rapid depletion of the oil due to the increasing number of vehicles, the pollutant emissions within its combustion products that threaten the ecological system and the concerns about the security of supply due to the oil reserves unevenly distributed over the globe, of which about 50% is located in the Middle East, encourages the exploration of fuel sources that are more environmentally friendly and have widespread reserves in the world [1].

Gasoline and diesel fuels that are produced from crude oil can also be produced synthetically from CO and H_2 gases with the method found by the German chemists Franz Fischer and Hans Tropsch in 1923. Fischer-Tropsch synthesis, a patented method since 1926, provides obtaining synthetic liquid fuel from many different kinds of carbon and hydrogen-derived raw materials. Generally, coal, natural gas and methane are used to obtain large amounts of CO and H_2 gases that are necessary for synthesis reactions. Today, Germany, India, China and South Africa that have major coal reserves produce commercially synthetic fuels with Fischer-Tropsch synthesis [2–4]. However, because the compositions of the synthetic gasoline and diesel fuels are similar to the natural gasoline and diesel fuels, their effects on the pollutant emissions resulting from vehicles are also similar.

In this chapter, for the purpose of reducing pollutant emissions resulting from internal combustion engines, the characteristics of hydrogen, natural gas, acetylene and ethanol, which are alternative fuels and can be used without requiring a structural change in SI and CI engines, and their effects on engine performance and exhaust emissions are mentioned. The physical and chemical characteristics

Properties	Acetylene	Hydrogen	CNG	Ethanol	Gasoline	Diesel
Formula	C_2H_2	H_2	CH_4	C_2H_5OH	C_4–C_{12}	C_8–C_{20}
Density (1 atm, 20°C (kg/m^3))	1.092	0.08	0.65	809.9	720–780	820–860
Auto ignition temperature (°C)	305	572	540	363	257	254
Stoichiometric ratio (kg/kg)	13.2	34.3	17.2	9	14.7	14.5
Motor octane number	45–50	130	105	89.7	95–97	–
Flammability limits in air (%Vol.)	2.5–81	4–74.5	5.3–15	3–19	1.4–7.6	0.6–5.5
Adiabatic flame temperature (K)	2500	2400	2320	2193	2300	2200
Min. quenching diameter (mm)	0.85	0.9	3.53	2.97	2.97	–
Min. ignition energy (MJ)	0.019	0.02	0.29	0.23	0.23	–
Maximum flame speed (m/s)	1.5	3.5	0.42	0.61	0.5	0.3
Lower heating value (kJ/kg)	48.225	120.000	49.990	26.700	43.000	42.500

Table 1.
Physical and combustion properties of fuels [146–153].

Fuels	Resource	Expended energy [MJ/MJ fuel]	Greenhouse emissions [g CO_2/MJ]
Gasoline	Crude oil	0.18	13.8
Diesel	Crude oil	0.20	15.4
Natural gas	EU-mix NG	0.17	13.0
	Imported NG 7000 km	0.29	22.6
	Imported NG 4000 km	0.21	16.1
	LNG*	0.28	19.9
	Shale gas	0.10	7.8
	Synthetic from wind electricity	1.05	3.3
Ethanol	Sugar*	1.20	28.4
	Wheat*	1.31	55.6
	Other*	1.66	41.4
Hydrogen	Natural Gas*	1.10	118
	Coal*	1.45	237
	Biomass*	1.05	14.6
	Electricity*	3.11	190

*Average value.

Table 2.
Energy and greenhouse gas balance in WTT analyses for EU (2010–2020+) [154].

of gasoline, diesel fuel and alternative fuels that are mentioned in this chapter are shown in **Table 1**.

Fuels used in ICE are generally produced from primary resources. To convert a source to a fuel and bring this fuel to a vehicle, well to tank (WTT) analyzes are made in terms of energy consumption and greenhouse gas emissions. The energy and greenhouse gas balances obtained from WTT analyses based 2010–2020+ years for the alternative fuels in EU are shown in **Table 2**. When **Table 2** is investigated according to fuel types, the maximum energy is consumed for the production of hydrogen gas and the minimum energy is expended for gasoline fuel. On the other hand when **Table 2** has been compared in terms of resources, the highest energy consumption is obtained as 3.11 MJ/MJ by the using of electrolysis in hydrogen production, while the lowest energy consumption occurs as 0.1 MJ/MJ in the producing of shall gas removed from EU geography. It is seen from **Table 2** that the highest CO_2 value is produced in the obtaining of hydrogen gas and the least emission value is emitted for gasoline fuel. In terms of resources, while the highest greenhouse emissions value is obtained as 237 g CO_2/MJ in hydrogen production from coal, the lowest greenhouse gas produce is 3.3 g CO_2/MJ in the producing of synthetic natural gas from wind electricity.

2. Acetylene

Acetylene was used as fuel in internal combustion engines in the early 1900s. Gustave Whitehead used a 15 kW engine powered by acetylene on his flying machine in 1901. Towards the year 1940s, acetylene began to be used in automobiles. In those years, about 4000 licenses for the conversion of vehicles to alternate fuels had been issued, and more than half of them were for conversion to acetylene [5]. Nowadays acetylene is only used in metal and chemical industries and it is not used in vehicles. Nevertheless, experimental studies on the use of acetylene in ICE have gained momentum in recent years due to high flame speed and energy density.

Acetylene was first discovered by Edmund Davy in 1836. But thereafter it was forgotten. Marcellin Berthelot rediscovered this hydrocarbon compound in 1860. He coined the name "acetylene" to this compound [6].

Acetylene, the first member of the alkynes (C_nH_{2n-2}), is a colorless and odorless gas but with an odor similar to garlic if produced from calcium carbide. Acetylene gas does not occur in quantities in nature but it is commonly obtained from the reaction of calcium carbide with water [7]. Calcium carbide (CaC_2) is produced by heating the mixture of quicklime and coke in electric arc furnaces to 2000–2100°C. Quicklime (CaO) is produced by heating calcium carbonate ($CaCO_3$) about at 900°C. **Figure 1** shows a schematic representation of an integrated facility for the production of calcium carbide [8]. Moreover, the processes are seen in Eqs. (1) and (2) [8–10].

$$CaCO_3 + heat \rightarrow CaO + CO_2 \tag{1}$$

$$CaO + 3C \rightarrow CaC_2 + CO \tag{2}$$

Acetylene has higher flame speed and energy density than gasoline and diesel [11] hence acetylene engines could more approach thermodynamically ideal engine cycle efficiency. But the octane number of acetylene is lower than other fuels which use in internal combustion engines [12]. Therefore the maximum amount of acetylene consumption is limited to the onset of knock. Lower ignition energy, high flame speed, wide flammability limits and lower octane number leads to premature ignition and undesirable combustion phenomenon called knock [13, 14]. These are the main problems encountered in using acetylene as a fuel in internal combustion engines.

In SI engines, acetylene and gasoline are either injected into the intake manifold or directly into the cylinder and the mixture is ignited by spark plug at the end of the compression stroke. In diesel engines, acetylene is either inducted along

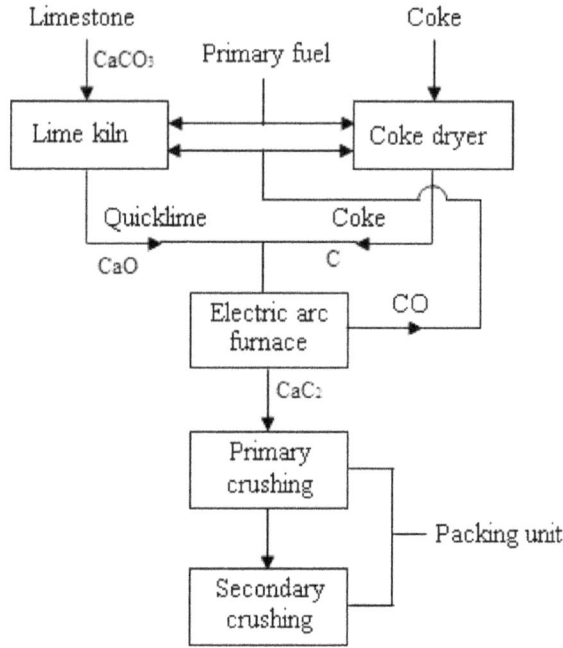

Figure 1.
Integrated calcium carbide production facility [8].

Load (%)	Gasoline (g/h)	Acetylene (g/h)	Acetylene (%)	Peak Pressure (bar)	Spark Advance (CA BTDC)
25	1877	0	0	16.6	21
	1320	500	27.5	16.5	13
	840	1000	54.3	15.6	2
50	2805	0	0	25.4	18
	2145	500	18.9	26.5	11
	1800	1000	35.7	20.9	1
75	3730	0	0	31.5	15
	3250	500	13.3	24.6	3
	2750	1000	26.7	23.8	0
100	4265	0	0	40.6	11
	3890	500	11.6	30.9	1
	3390	1000	22.8	29.0	-2*

*2 CA After Top Dead Center

Table 3.
Mass flows of fuels, peak pressure and spark advance [18].

with intake air or injected directly into the cylinder and compressed. However, the mixture of acetylene-air does not auto-ignite due to its very high self-ignition temperature. A small amount of diesel fuel called pilot fuel is injected into the mixture towards the end of the compression stroke. The pilot diesel fuel auto-ignites first and ignites the acetylene-air mixture such as spark plug. So, dual fuel diesel engines combine the features of both SI and CI engines [15–17].

The main advantages of using acetylene as gasoline-acetylene mixtures in SI engines [5, 18–21]:

- Acetylene-gasoline mixtures can be used in SI engines at every load from low load to full load. However, it can be also used as a single fuel at partial loads.

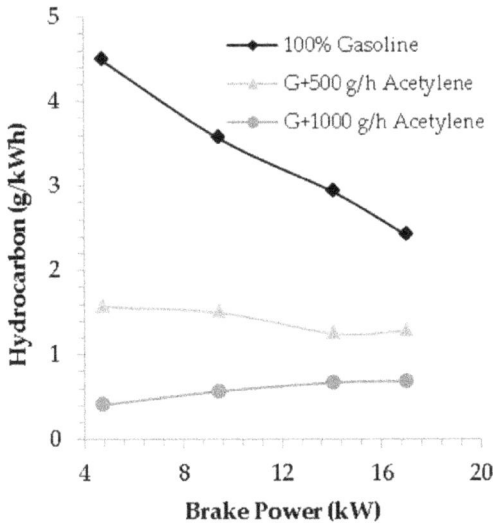

Figure 2.
Variety of HC with brake power (1500 rpm, different loads) [18].

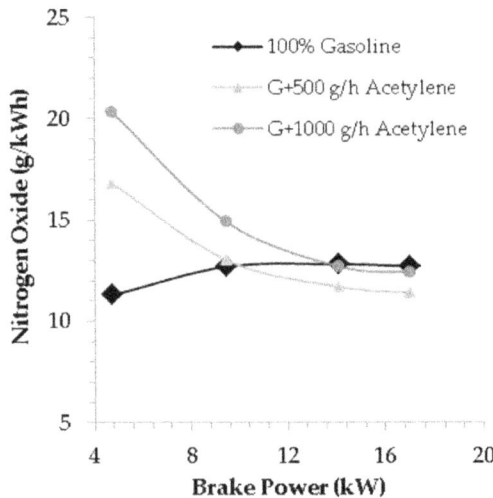

Figure 3.
Variety of NO with brake power (1500 rpm, different loads) [18].

- If acetylene is mixed with gasoline under stoichiometric conditions, it causes a decrease in gasoline consumption at constant output power as seen in **Table 3**. At the same time, as can be seen **Figure 2**, hydrocarbon emissions were significantly reduced at all loads and as can be seen **Figure 3**, NO emissions were reduced at full loads according to working with gasoline [18]. Experimental studies [18] were realized at 1500 rpm and stoichiometric ratio under 25, 50, 75% and full load conditions. The acetylene was injected into the intake manifold of test engine through the gas injector 500 and 1000 g/h gas flow rates.

- Acetylene increases the poor combustion limit in partial loads in SI engines. The engine can be operated in leaner conditions with gasoline-acetylene mixtures. As seen in **Figures 4** and **5** the brake thermal efficiency of the engine increases and the specific fuel consumption decreases. Further, at high equivalence ratios, the fairly reduced exhaust emissions are observed. NO emissions are almost non-existent as in-cylinder temperatures decrease in lean fuel-air

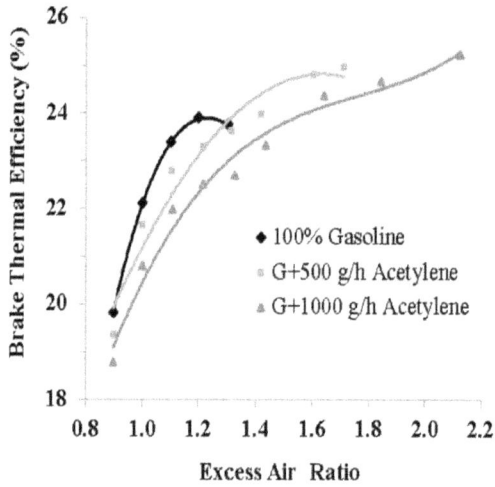

Figure 4.
The variation of BTE with excess air ratio (1500 rpm, 25% load) [19].

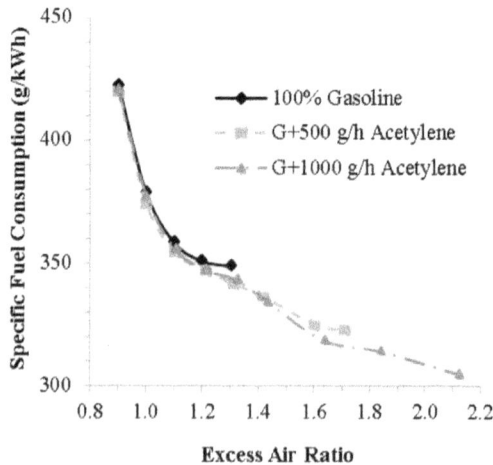

Figure 5.
The variation of BSFC with excess air ratio (1500 rpm, 25% load) [19].

mixtures and unburned hydrocarbon emissions are quite reduced when compared gasoline operation in SI engines as can be seen **Figures 6** and 7. With the
use of acetylene as an alternative fuel in SI engines, air pollution from SI engine
vehicles in large cities can be significantly reduced [19].

- Acetylene operates in diesel engines with dual fuel mode by a little engine
  modification and while reduces NOx, HC, CO and CO_2 emissions, contributing
  to a significant reduction in diesel fuel consumption [16]. Acetylene cannot
  be used as a single fuel in diesel engines due to the high compression ratio. In
  that study, the tests were conducted on a four-stroke diesel engine with a rated
  power output of 4.4 kW at 1500 rpm, with slight modification in intake manifold for holding the gas injector. The gas flow rates of 110, 180 and 240 g/h
  and optimized injection timings were arranged by ECU's. **Table 4** gives energy
  share ratio of diesel and acetylene at 240 g/h flow rate [16].

- In countries with large coal reserves and little or no oil reserves acetylene can
  be used in automobiles that form the largest part of vehicle traffic. Thus the
  country's need for oil can be reduced.

The main disadvantages of acetylene as alternative motor fuel [22–26]:

- Acetylene is a very explosive gas which sensitive to pressure and temperature.
  For this reason, in vehicles that use acetylene as fuel should be security precau-
  tions taken seriously and should not be parked in closed areas.

- Acetylene is a fuel with very low ignition energy and may cause backfire in
  intake manifold.

- As the knock resistance of acetylene is low, the air-fuel ratio must be precisely
  adjusted to avoid knock.

- Acetylene can be used as the only fuel in SI engines only under very lean
  air-fuel mixture conditions. In very lean conditions, we cannot get maximum
  power out of the engine.

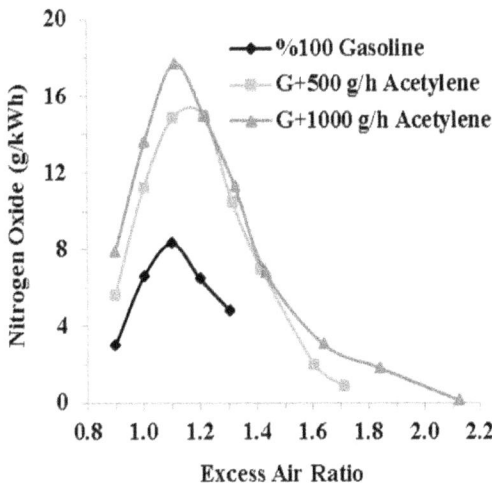

Figure 6.
The variation of NO with excess air ratio (1500 rpm, 25% load) [19].

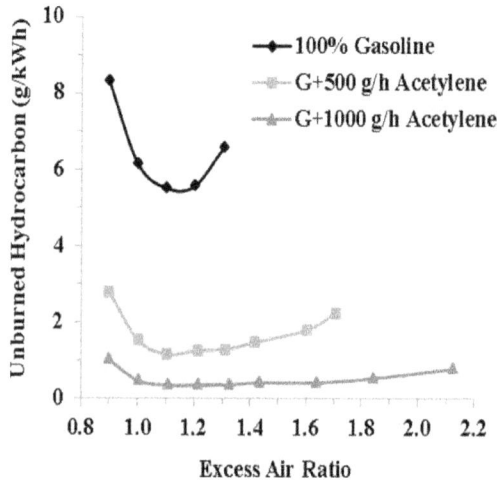

Figure 7.
The variation of UHC with excess air ratio (1500 rpm, 25% load) [19].

Load (%)	Energy equivalent of diesel fuel (kW)	Energy equivalent of acetylene fuel (kW)	Energy share of gas (%)	Energy share of diesel (%)
0	4.01	3.21	44	56
25	5.31	3.21	38	62
50	7.79	3.21	29	71
75	9.33	3.21	26	74
100	10.39	3.21	24	76

Table 4.
Energy share ratio of diesel and acetylene at 240 g/h [16].

- Storage of acetylene in vehicles is an unsolved problem yet. As acetylene is decomposed at a pressure of 2.5 bar, it cannot be stored as compressed gas like other gases. Acetylene is stored dissolved in acetone contained in a metal cylinder with a porous filling material under 18 bar pressure. When acetylene cylinders are empty, on-site filling is not possible. Therefore, disassembly and montage the cylinder is a major disadvantage. Although manufactured in different sizes, cylinders that can be stored 8.7 m³ acetylene have a volume of about 60 liters and average weighs (full) 70 kg [27]. This situation causes great difficulties in practice.

- Another method is to produce acetylene from carbide as in the 1940s and to use it without storage. This method requires a complex system as shown in **Figure 1**. Disposal of the residue called calcium hydroxide is another important problem of an on-board fuel generating system

3. Natural gas

Natural gas is a fossil fuel found in nature reserves, associated or not with petroleum [28]. The cost of obtaining from nature is lower than other fossil fuels. Natural gas consists of about 90% methane, 3% ethane, 3% nitrogen, 2% propane and other trace gases. Methane which is the always dominant component of natural gas is the first member of alkane's family. Since it has a high H/C ratio, natural gas is known as the cleanest fuel in fossil fuels. Due to its ecological benefits, city buses operate with natural gas engines in many countries. CO_2 gas, which should normally be between 180 and 280 ppm in the atmosphere, reached 405 ppm as of September 2018 due to overuse of fossil fuels [29]. Therefore, many countries encourage the use of natural gas instead of petrol and diesel fuel in vehicles. Because, natural gas mixes perfectly with air, it is easy to ignite, provides clean combustion and gives high heat. The thermal efficiency of natural gas engines is higher than that of gasoline engines due to these engines have a higher compression ratio than gasoline engines [28–35].

Unlike gasoline and diesel engines, natural gas-powered internal combustion engines do not require fuel enrichment in cold start, and exhaust emissions are not affected by low temperatures. Natural gas vehicles (NGV) produce emission values lower than the EURO 6 standard according to vehicles using petroleum-derived fuel [30].

According to NGV Global's report, the number of NGV and filling stations in the world is increasing rapidly (**Figures 8** and **9**). China ranks first in the NGV Park with 6,080,000 vehicles and 8400 filling stations, according to 2018 data. In the number of NGV, Iran, India and Pakistan are the countries that come after China. The total number of NGVs reached to 26,130,000 as of June 2018 [31].

The biggest disadvantage for the NGV transportation sector comes from the storage challenge of natural gas. Natural gas is a lighter gas than air. While the density of air at sea level at 15°C is 1.225 kg/m³, although the density of natural gas varies according to its composition, it is about 0.71 kg/m³. As natural gas is a light gas the energy density per unit volume is low and in order to ensure a reasonable driving distance the storage volume should be chosen large. Fortunately, technology has developed and the natural gas has been begun to storage in steel or carbon tubes at a pressure of 200 bar with high pressure compressors. Parking of natural gas vehicles in enclosed spaces is dangerous for safety reasons. Nowadays, cars with natural gas engines have a range of more than 300 miles with a single filling. Also, natural gas is not a renewable energy source, like other fossil fuels [35–37].

High knock resistance of natural gas allows it to be used in engines with higher compression ratios as compared to gasoline engines. Operation of natural gas vehicles at higher compression ratios than gasoline vehicles increases the thermal

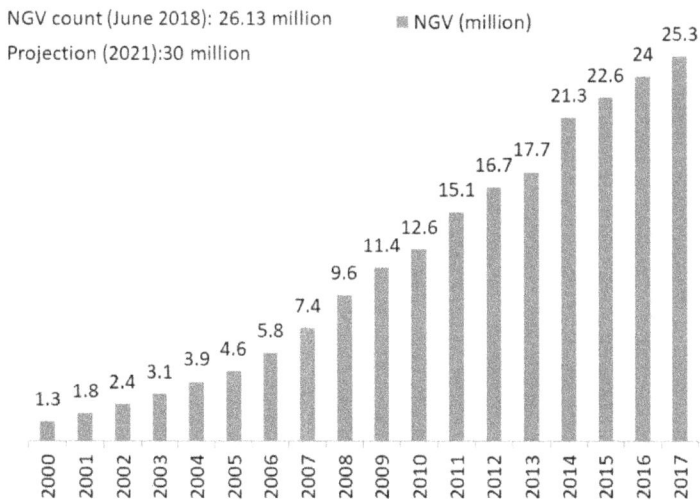

NGV count (June 2018): 26.13 million

Projection (2021):30 million

NGV (million)

Figure 8.
Number of natural gas vehicles worldwide by years [31].

efficiency. As seen in **Figure 10**, in the tests carried out at different compression ratios with natural gas and natural gas-hydrogen mixtures (HCNG), the minimum fuel consumption for the compression ratio of 12.5 was obtained. **Figure 11** shows that, THC emissions are lower than the Euro VI standards in all compression ratios [30]. The experiments have been carried out using a modified diesel engine having 9.6, 12.5 and 15 different compression ratios at 1500 rpm under full load conditions fueled by hydrogen enriched compression natural gas blends (100% CNG, 95% CNG + 5% H$_2$, 90% CNG + 10% H$_2$ and 80% CNG + 20% H$_2$). Engine performances

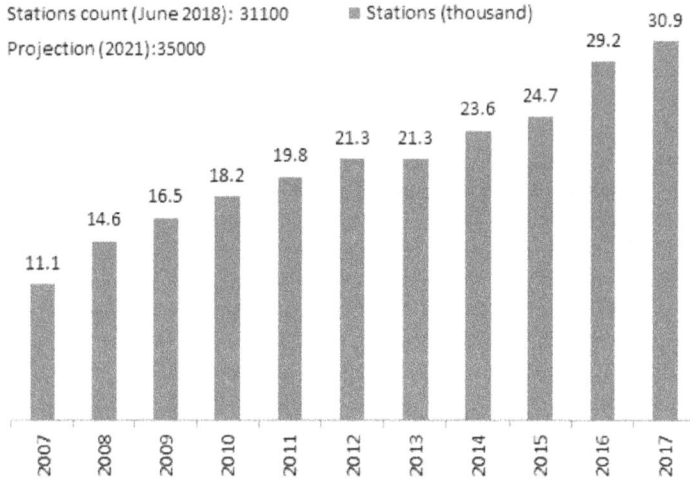

Figure 9.
Number of natural gas fueling stations worldwide by years [31].

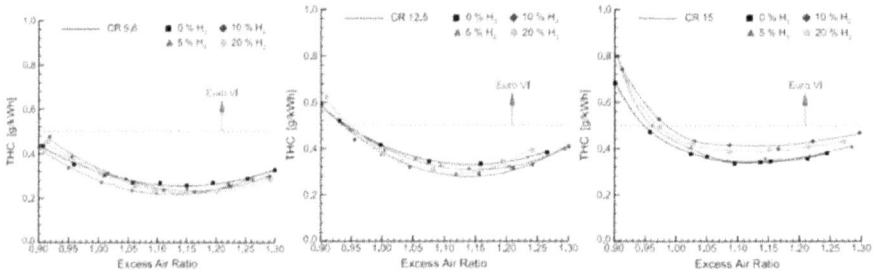

Figure 10.
The THC values versus excess air ratio using different compression ratios [30].

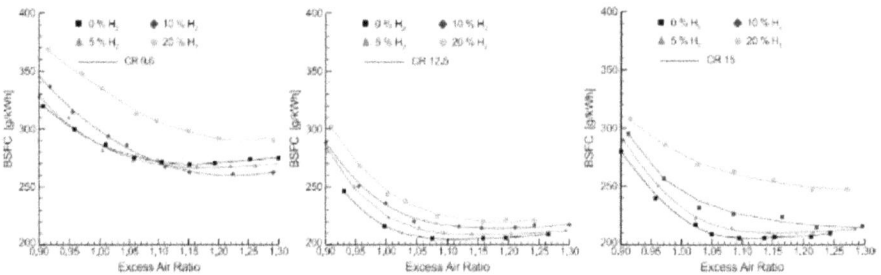

Figure 11.
The BSFC values versus excess air ratio using different compression ratios [30].

CR	H_2 (%)	$\lambda = 1.0$	$\lambda = 1.15$
9.6	0	2000	3620
	5	2100	3825
	10	1710	4185
	20	1535	4225
12.5	0	2040	4410
	5	1940	4200
	10	2260	4520
	20	2210	4695
15	0	2045	4465
	5	2570	4700
	10	2660	4565
	20	3030	4350

Table 5.
NO_X values (ppm) for $\lambda = 1.0$ and $\lambda = 1.15$ [30].

and emissions parameters have been realized at 10°CA BTDC ignition timing and different excess air ratios ($\lambda = 0.9$–1.3).

NO_X values for $\lambda = 1.0$ and $\lambda = 1.15$ show in **Table 5**. As seen in the table, increasing of compression ratio and hydrogen fraction values lead to an increase in NO_X values.

4. Ethanol

Ethanol is generally produced from renewable sources such as biomass and agricultural feedstock [38, 39]. So, ethanol has been used widely as alternative fuel in internal combustion engines. The octane number of ethanol is higher than the octane number of the gasoline. The high octane number of ethanol allows the use of ethanol as fuel in an SI engine with a higher compression ratio [40]. The latent vaporization heat of ethanol increase cooling effect in the cylinder, this situation leads to an increase in volumetric efficiency [41]. Ethanol burns cleaner than gasoline and diesel fuels and it produce less CO, CO_2 and NO_x. It has low diffusivity and ignition difficulty at low temperature, therefore combustion is not completed at low temperature and HC increases compared to gasoline in ethanol use. Ethanol chemical formulation is C_2H_5OH. Hydrogen percentage of ethanol is higher than gasoline.

Recently environmental authorities in large urban centers have expressed their concerns on the true effect of using ethanol blends of up to 20% in-use vehicles without any modification in the setup of the engine control unit (ECU), and on the variations of these effects along the years of operation of these vehicles [40].

Pure ethanol can be used internal combustion engines but there are some problems [42–45]. These problems are;

1. Ethanol has a low flame speed. So it has a bad cold-starting function. The using as fuel is hard in the winter months.

2. There is no passenger car designed for 100% ethanol. The use of pure ethanol can damage engines. Even engines that can work with gasoline-ethanol mixtures can reach up to 85% ethanol.

3. Ethanol is a corrosive fuel. So, the materials and surfaces of parts of combustion chamber, all plastic materials having contact with fuel and fuel injection system must be improved.

5. Hydrogen

Although hydrogen the most common element in the world and it does not exist in nature in its pure state, so it has to be produced from sources like water and natural gas. The environmental impact and energy efficiency of hydrogen depends on how it is produced [46, 47].

Hydrogen has been studied as an alternative gas fuel for a long time. Hydrogen has not some problems associated with liquid fuels, such as vapor lock, cold wall quenching, inadequate vaporization and lean mixing. Hydrogen has clean burning behaviors. As hydrogen is burned, it products mainly water. The combustion of hydrogen does not bring out toxic products such as hydrocarbons, carbon monoxide and carbon dioxide [48]. The most important advantage of hydrogen is that it does not produce CO_2 gas, which is one of the most important sources of global warming. In addition, hydrogen has a wider limit of flammability than gasoline, diesel and natural gas [49, 50]. Moreover, hydrogen has high flame speed and it has high self-ignition temperature [51]. Also, hydrogen can easily burn in ultra-lean mixtures [52]. The energy required to ignite the hydrogen-air mixture is only 0.02 MJ. Therefore, it is ideal for poor mixed burns [50]. Finally, hydrogen can be used at wide compression rates in internal combustion engines as the self-ignition temperature of hydrogen is too high [53]. Due to these properties, many studies have been carried out on the use of hydrogen in internal combustion engines [54–56].

Due to the low energy required for the ignition of hydrogen, the mixture immediately ignites when it comes into contact with a hot spot in the cylinder. As a result, knock may occur [56, 57]. As can be seen from **Figure 12**, another disadvantage of hydrogen is its low energy density [58]. In addition, the formations of NO_X emissions are increased by hydrogen combustion due to high flame temperature [59, 60]. The increasing of NO_X with hydrogen can be seen from **Figure 13**.

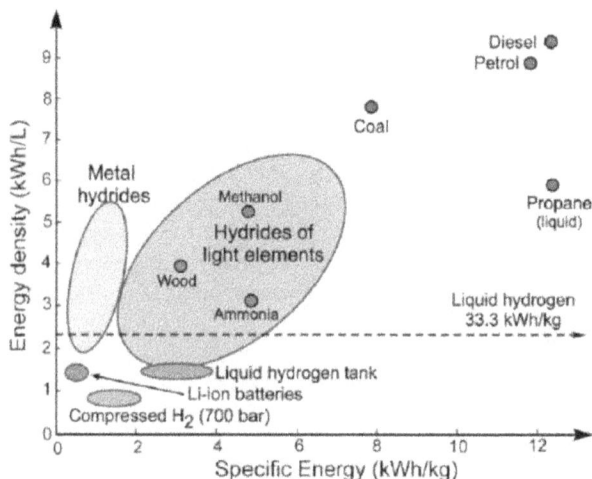

Figure 12.
The energy density of some fuels [145].

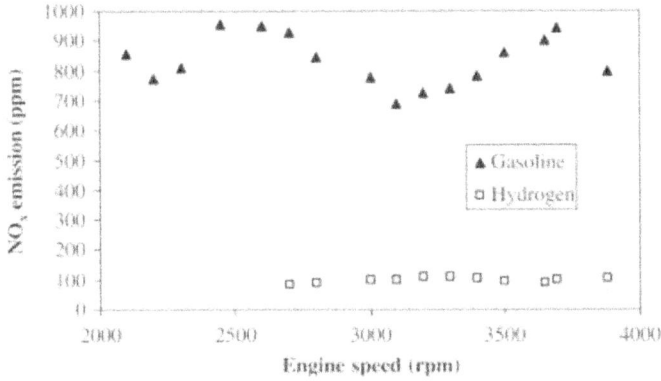

(a)

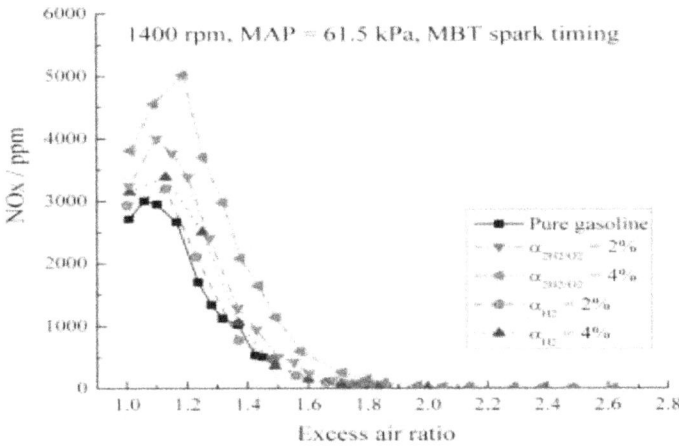

(b)

Figure 13.
NO$_X$ variations at different engine speeds (a) [61] and different excess air ratio (b) [62] with addition of hydrogen to gasoline.

The experiments in the study fueled by pure hydrogen and gasoline [61], in which **Figure 13** was taken, carried out on a four-cylinder, four stroke, SI engine with carburetor, having 8.8:1 compression ratio. The ignition timing was set to 10° before top dead center (BTDC). The engine was run between 2600 and 3800 rpm engine speeds. In the experimental study [62], the tests were carried out at 1400 rpm engine speed, 61.5 kPa manifold air pressure, MBT spark timing and different excess air ratios (1.0–2.6). In this study, to simulate the hydroxygen, the hydrogen-to-oxygen mole ratio was fixed at 2:1 through adjusting the injection durations of hydrogen and oxygen. Moreover, three standard hydroxygen volume fractions in the total intake gas of 0, 2 and 4% were adopted in the tests.

6. Hydrogen mixture

Because of hydrogen has some negative effects on internal combustion engine, it is used as a mixture rather than pure. The most widely mixture of hydrogen is

HCNG. The mixture has been formed by the blending of natural gas. Natural gas-hydrogen mixtures (HCNG), which are considered as alternative fuels for conventional engines, are mixtures formed to combine the superior properties of natural gas and hydrogen. There are many studies [63–70] using HCNG as an alternative fuel.

As can be seen in **Figure 14**, the hydrogen adding causes an increase in thermal efficiency and causes an expansion of the flammability limits. In addition, when the figures are examined, it is seen that the addition of hydrogen increases the stability of combustion and the value of brake power and reduces the specific fuel consumption.

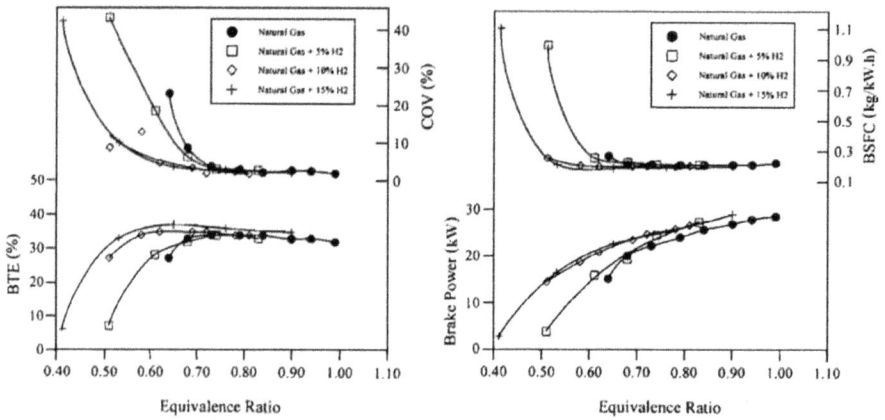

Figure 14.
BTE, COV, power and BSFC values versus equivalence ratio at 2200 rpm, 50% WOT with MBT timing and different hydrogen percent [69].

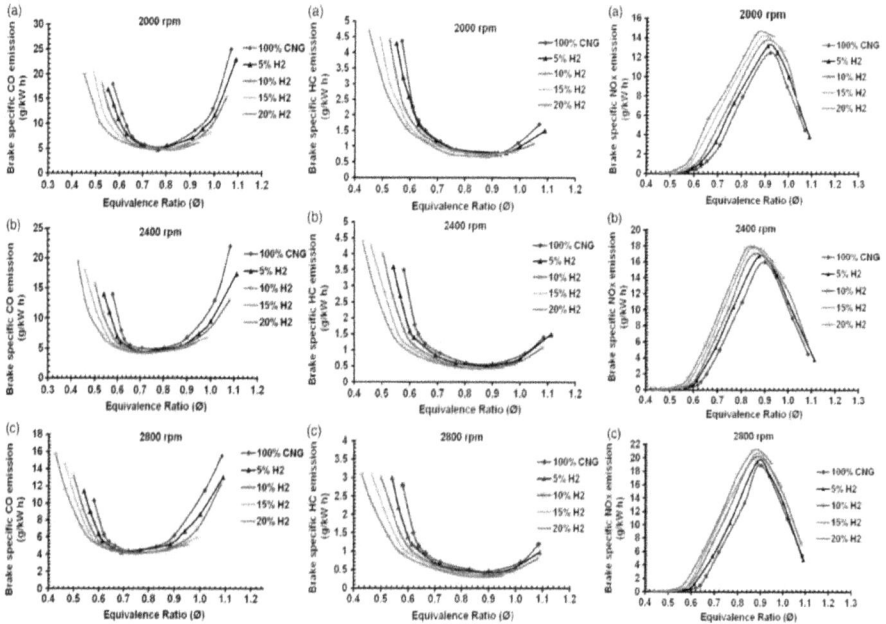

Figure 15.
Emission values versus equivalence ratio at 2000 rpm (a), 2400 rpm (b) and 2800 rpm (c) and different hydrogen rates [70].

Moreover, as can be seen in **Figure 15**, the addition of hydrogen to natural gas leads to a decrease in CO and HC emissions and an increase in NO_X values. In the experimental study, in which **Figure 15** was taken, the experiments were performed at 2000, 2400 and 2800 rpm with wide open throttle and varying the equivalence ratio. The engine with single-cylinder having 7.25:1 compression ratio was fueled by compressed natural gas, and mixtures of hydrogen in CNG as 5, 10, 15 and 20% by energy.

Another mixture made using hydrogen is the ethanol-hydrogen mixture. In the literature, it can be found many studies on the use of hydrogen and ethanol in internal combustion engines [71–85].

In the experimental study [85], in which **Figure 16** was taken, the experiments were carried out on a compression ignition engine modified to run on spark ignition mode fueled with hydrogen-ethanol dual fuel combination with different percentage of hydrogen (0–80%) under compression ratio conditions of 7:1, 9:1 and 11:1 by varying the spark ignition timing at a constant speed of 1500 rpm.

In a study conducted with a mixture of hydrogen-acetylene, Sampath Kumar et al. [86] have been investigated the performance and the emission behaviors of SI

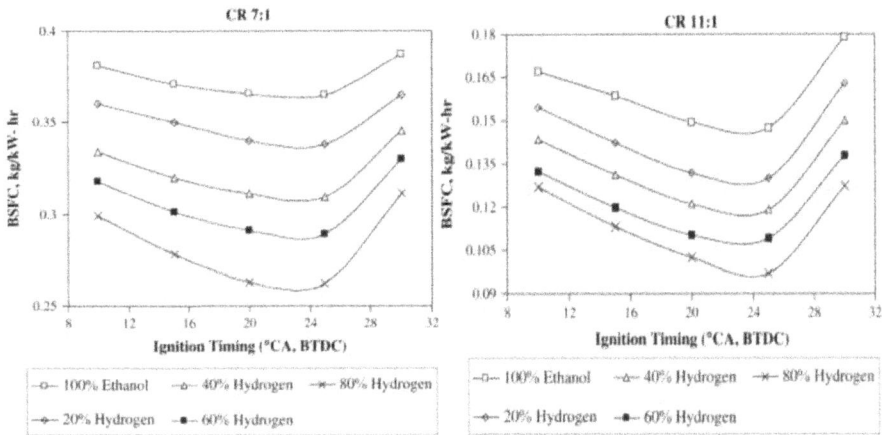

Figure 16.
The BSFC variations versus ignition timings at 7:1 and 11:1 compression ratios for different ethanol-hydrogen blend [85].

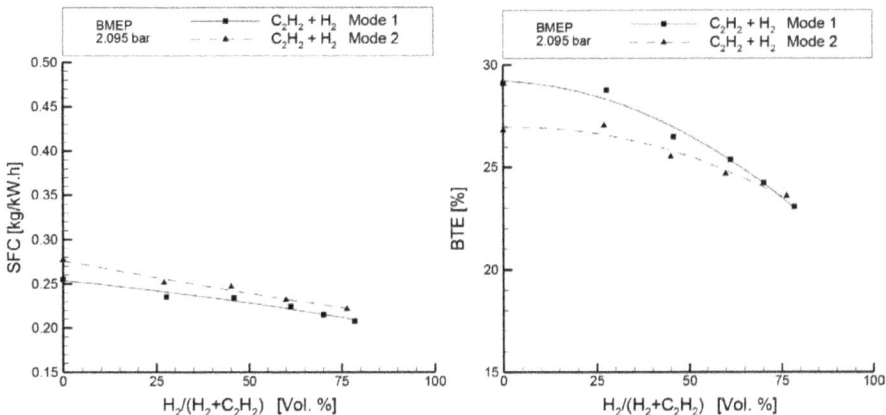

Figure 17.
SFC and BTE values versus different fractions of hydrogen [87].

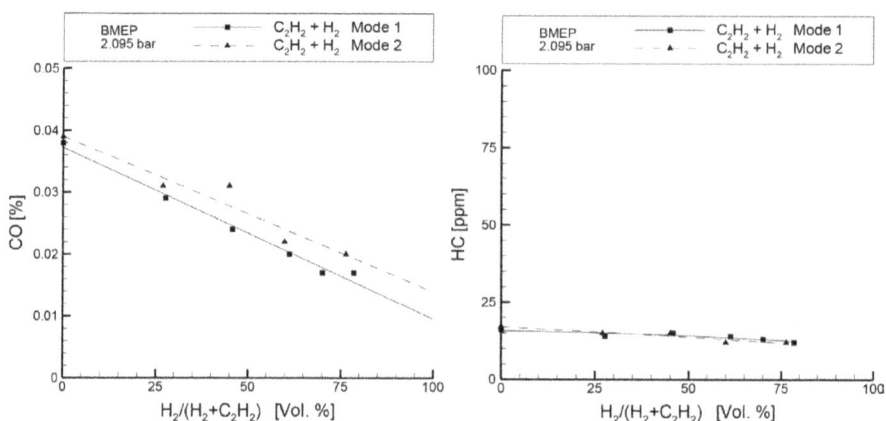

Figure 18.
CO and HC emissions versus different fractions of hydrogen [87].

engine fueled by hydrogen-acetylene fuel. The results indicated that brake thermal efficiency raised and emissions values descended when compare to gasoline.

In the another study, Tangöz et al. [87] have been analyzed the performance and emission values of an SI engine fueled by acetylene-hydrogen at a fixed BMEP value of 2.095 bar, a load of 30 Nm and an engine speed of 1500 rpm under lean mixture conditions (λ = 1.3–2.8). As can be seen from **Figures 17** and **18**, the experimental results showed that the values of specific fuel consumption are declined between 18.5 and 20.1% by hydrogen addition in the blend. The values of brake thermal efficiency are declined between 6.2 and 3.3% with the addition of hydrogen in the blend. The curves of cylinder pressure and heat release rate are advanced to top dead center by the adding of hydrogen to acetylene. The adding of hydrogen in acetylene leads to a decrease in CO and HC emissions and an increase in NO_x values for fixed lambda.

7. Alternative fuels for new ICE applications

Today, one of the most important problems in the use of internal combustion engines is the production of harmful emission gases. For this reason, many studies have been carried out to reduce the emissions while maintaining of engine performances, with the new ICE applications such as HCCI, RCCI, PCCI and PPC. Moreover, for the purpose of reducing emissions, some of these studies focused on the using of alternative fuels. In the new engine applications have a process in which a homogeneous mixture of air and fuel is compressed under the conditions in which auto ignition occurs close to the end of the compression stroke, followed by combustion, which is significantly faster than conventional diesel or Otto combustion. The auto-ignition and combustion phasing in cylinder are controlled by mixture stratification and fuel injection timing [88–93]. These engine applications compared to conventional engines allows to reduce nitrogen oxide and soot emissions and to achieve higher thermal efficiency [94–98]. However, it is very difficult to control the auto ignition in these engines. Many studies were carried out to control the auto ignition process in the engines by using alternative fuels having high auto ignition temperature or low reactivity or high octane number.

One of the most important new ICE applications is homogeneous charge compression ignition (HCCI). To control the auto ignition process in HCCI engine, some fuels having high auto ignition temperature use as alternative fuel. When these studies are examined, it is seen that the studies focused on the natural gas [99–104], ethanol [105–108], acetylene [109–114] and hydrogen [115–122]. Reactivity controlled compression ignition (RCCI), premixed charge compression ignition (PCCI) and partially premixed combustion (PPC) are other new ICE applications. In the engine applications, the low reactivity fuel is introduced from port injection to form a homogeneous mixture in the cylinder, and the high cetane number fuel is injected directly into the cylinder to control the combustion phasing and duration. High octane fuels or low reactivity with resistance to spontaneous ignition are more favorable for RCCI, PCCI and PPC combustion. For this reason, most of the studies carried out on RCCI, PCCI and PPC engines are focused on natural gas [89, 123–133] and ethanol [134–144] as an alternative fuel.

As a result, the operation parameters such as fuel type, fuel composition, air fuel ratio and inlet temperature were observed to significantly affect working regime of the new ICE applications. However, it is considered that a complete framework for the each ICE application modes has been not provided. Moreover, in spite of significant reduction in NO_X and soot emissions is observed in the applications fueled by the alternative fuels, significant amounts of HC and CO emissions forming still remain problematic.

8. Conclusion

Acetylene has some suitable properties such as high energy density, high flame temperature, high flame speed and low emission production. For this reason, it is considered to can be use an important contribution fuel or alternative fuel in the future for internal combustion engine. It increases brake thermal efficiency while contribute to decrease fuel consumption and all emission values. However, some studies should be carried out to increase the knock resistance of acetylene. Moreover, efficient production methods and new storage methods need to be developed in order to use acetylene as an alternative fuel in vehicles. Finally, in order to determine whether acetylene is economical or not, well to tank analysis should be performed.

Looking at today's applications, it is seen that natural gas fuel is a suitable fuel especially for SI engines having high compression ratio due to high knock resistance. Operation of natural gas vehicles at higher compression ratios than gasoline vehicles decreases the BSFC. On the other hand, natural gas, the cleanest fossil fuel due to having high H/C ratio, provides more reduction in THC emission values than Euro VI standard when suitable compression ratio is met. However, the storage problem must be eliminated in order to be used in all engines. Moreover, studies should also be done to increase the energy density.

Ethanol has high octane number. However, it is expensive than fossil fuels and it has corrosive property. In addition, even engines that can work with gasoline-ethanol mixtures can reach up to 85% ethanol. Ethanol can be blended to other alternative fuel to improve the energy density. Ethanol burns cleaner than gasoline and diesel fuels and produces less CO, CO_2 and NO_x but HC increases due to it has low diffusivity and ignition difficulty at low temperature.

Hydrogen is a clean fuel and the mass energy density is very high. Fast burning characteristics of hydrogen permits high speed engine operation and less heat loss occurs for hydrogen than gasoline. NO_x emission of hydrogen fuelled engine is about 10 times lower than gasoline fuelled engine if it works lean conditions. Because of hydrogen has some disadvantages such as very low ignition energy and volume energy density, it is mixed with other fuels especially natural gas to use in SI engines.

Intensive studies such as the use of hydrogen in a liquid state should be done to solve the storage problems in order to achieve the desired level of use in internal combustion engines. Also, the methods or mixtures that reduce NO_x formation should be studied.

In spite of significant reduction in NO_X and soot emissions is observed in the new ICE applications such as HCCI, RCCI, PCCI and PPC fueled by the alternative fuels, significant amounts of HC and CO emissions forming still remain problematic.

Consequently, each fuel has positive and negative properties for use in internal combustion engines. There are differences in the effects of each alternative fuel on emissions and engine performance. The future studies could be carried out to obtain an appropriate hybrid fuel by making a comparison between these alternative fuels to reduce all emissions and to improve engine performance.

Abbreviations

BMEP	brake mean effective pressure
BSFC	brake specific fuel consumption
BTE	brake thermal efficiency
CA BTDC	crank angle before top dead center
CI	compression ignition engine
COV	coefficient of variation
CR	compression ratio
EU	European Union
HCNG	natural gas-hydrogen mixtures
ICE	internal combustion engine
MBT	maximum brake torque
NGV	natural gas vehicles
SI	spark ignition
WOT	wide open throttle
WTT	well to tank

Author details

Mehmet Ilhan Ilhak[1], Selim Tangoz[2], Selahaddin Orhan Akansu[1*] and Nafiz Kahraman[3]

1 Department of Mechanical Engineering, Faculty of Engineering, Erciyes University, Kayseri, Turkey

2 Department of Airframes and Powerplants, Faculty of Aeronautics and Astronautics, Erciyes University, Kayseri, Turkey

3 Department of Astronautical Engineering, Faculty of Aeronautics and Astronautics, Erciyes University, Kayseri, Turkey

*Address all correspondence to: akansu@erciyes.edu.tr

IntechOpen

References

[1] Türkiye Petrolleri. 2017 Yılı Ham Petrol ve Doğalgaz Sektör Raporu. 2018. 50 s

[2] Schulz H. Short history and present trends of Fischer-Tropsch synthesis. Applied Catalysis A: General. 1999;**186**(1-2):3-12

[3] Dry ME. High quality diesel via Fischer-Tropsch process—A review. Journal of Chemical Technology and Biotechnology. 2001;77:43-50

[4] Dry ME. The Fischer Tropsch process 1950-2000. Catalysis Today. 2002;**71**:227-241

[5] http://www.douglas-self.com/MUSEUM/POWER/acetylene-eng/acetyleneeng.htm [Accessed: 2018]

[6] https://todayinsci.com/D/Davy_Edmund/DavyEdmundBio.htm [Accessed: 2018]

[7] Odell WW. Facts relating to the production and substitution of manufactured for natural gas/acetylene (C_2H_2). Bulletin, 300-308. U.S. Government Printing Office. 1829. p. 64

[8] https://www3.epa.gov/ttn/chief/ap42/ch11/bgdocs/b11s04.pdf [Accessed: 2018]

[9] Kannan P, Viswabharathy P, Kumar PD. Experimental study of carbide as an alternate fuel using in internal combustion engine. International Journal of Emerging Technologies in Engineering Research. 2017;**5**(5):92-104

[10] Oršula I, Lehocký M, Steltenpohl P. Simulation of calcium acetylide and acetylene production. Acta Chimica Slovaca. 2015;**8**(2):91-96

[11] Behera P, Jha AK, Murugan S. Dual fuel operation of used transformer oil with acetylene in a di diesel engine. International Journal on Theoretical and Applied Research in Mechanical Engineering. 2013;**2**(2):126-132

[12] Kumar P, Reddy SJ, Bodukuri K. Internal combustion engine using acetylene as an alternative fuel. International Journal of Engineering Research and Development. 2018;**14**(5):56-61

[13] Lakshmanan T, Nagarajan G. Performance and emission of acetylene-aspirated diesel engine. Jordan Journal of Mechanical and Industrial Engineering. 2009;**3**(2):125-130

[14] Sudheer. Experimental performance analysis of acetylene aspirated diesel engine. International Journal for Scientific Research & Development. 2016;**4**(06):2321-0613

[15] Lakshmanan T, Nagarajan G. Experimental investigation on dual fuel operation of acetylene in a DI diesel engine. Fuel Processing Technology. 2010;**91**:496-503

[16] Lakshmanan T, Nagarajan G. Experimental investigation of timed manifold injection of acetylene in direct injection diesel engine in dual fuel mode. Energy. 2010;**35**:3172-3178

[17] Lakshmanan T, Nagarajan G. Experimental investigation of port injection of acetylene in DI diesel engine in dual fuel mode. Fuel. 2011;**90**:2571-2577

[18] Ilhak Mİ, Akansu SO, Kahraman N, Unalan S. Experimental study on an SI engine fuelled by gasoline-acetylene mixtures. Energy. 2018;**151**:707-714

[19] Ilhak Mİ. Investigation of the effect of acetylene gas on the engine performance and emissions in an SI engine [PhD thesis]. 2018. p. 156

[20] Hilden DL, Stebar RF. Evaluation of acetylene as a spark ignition engine fuel. International Journal of Energy Research. 1979;**3**:59-71

[21] Gupta K, Suthar K, Jain SK, Agarwal GD, Nayyar A. Design and experimental investigations on six-stroke SI engine using acetylene with water injection. Environmental Science and Pollution Research. 2018;**25**:23033-23044

[22] Shaik Khader Basha SK, Rao PS, Rajagopal K. Experimental investigation of performance of acetylene fuel based diesel engine. International Journal of Advancements in Technology. 2016;**7**(1):3-7

[23] Pravinkumar SC, Bhavsar AA. Experimental investigation of diesel engine operating parameters for a mixture of acetylene and turpentine oil with diesel by design of experiment. International Journal for Innovative Research in Science & Technology. 2017;**3**(2):11-16

[24] Sahu GK, Kumar S. Performance analysis of four stroke diesel engine working with acetylene and diesel. International Journal for Research in Applied Science & Engineering Technology. 2017;**5**(8):2038-2043

[25] Mahla SK, Kumar S, Shergill H, Kumar A. Study the performance characteristics of acetylene gas in dual fuel engine with diethyl ether blends. International Journal on Emerging Technologies. 2012;**3**(1):80-83

[26] Nataraja M, Kiran Kumar M, Manjunath K, Madhukumar K. Acetylene as an alternate fuel in modified 4-stroke spark ignition engine. International Journal of Innovative Research in Science, Engineering and Technology. 2018;**7**(7):351-360

[27] https://www.supagas.net.au/acetylene-8-7m3-cylinder.html#.W8QPAnszbIU [Accessed: 2018]

[28] Neiva L, Gama L. The Importance of Natural Gas Reforming, Natural Gas. InTech; 2010. ISBN: 978-953-307-112-1. Available from: http://www.intechopen.com/books/natural-gas/the-importance-of-natural-gas-reforming [Accessed: 2018]

[29] https://www.esrl.noaa.gov/gmd/ccgg/trends/monthly.html [Accessed: 2018]

[30] Tangöz S, Akansu SO, Kahraman N, Malkoç Y. Effects of compression ratio on performance and emissions of a modified diesel engine fueled by HCNG. International Journal of Hydrogen Energy. 2015;**40**:15374-15380

[31] http://www.iangv.org/current-ngv-stats/ [Accessed: 2018]

[32] Demirbas A. Chapter 2: Methane gas hydrate. In: Natural Gas. London, England: Springer; 2010. pp. 57-76

[33] Aljamali S, Mahmood WMFW, Abdullah S, Ali Y. Comparison of performance and emission of a gasoline engine fuelled by gasoline and CNG under various throttle positions. Journal of Applied Sciences. 2014;**14**:386-390

[34] Tabar AR, Hamidi A, Ghadamian H. Experimental investigation of CNG and gasoline fuels combination on a 1.7 L bi-fuel turbocharged engine. International Journal of Energy and Environmental Engineering. 2017;**8**:37-45

[35] Baranes E, Jacqmin J, Poudou JC. Non-renewable and intermittent renewable energy sources: Friends and foes? Energy Policy. 2017;**111**:58-67

[36] Kato K, Igarashi K, Masuda M, Otsubo K, Yasuda A, Takeda K. Development of engine for natural gas vehicle. 1999. SAE Paper No. 1999-01-0574.1-11

[37] Dubois LH. Adsorbed Natural Gas On-Board Storage for

Light-Duty Vehicles. California Energy Commission. 2017. Publication Number: CEC-500-2017-038

[38] Fulton L, Howes T, Hardy J. Biofuels for Transport—An International Perspective. Paris: International Energy Agency; 2004

[39] Zvirin Y, Gutman M, Tartakovsky L. Fuel effects on emissions. In: Sher E, editor. Chapter 16 in the Handbook of Air Pollution from Internal Combustion Engines, Pollutant Formation and Control. San Diego, USA: Academic Press; 1998. pp. 548-651

[40] Tibaquirá JE, Huertas JI, Ospina S, Quirama LF, Niño JE. The effect of using ethanol-gasoline blends on the mechanical, energy and environmental performance of in-use vehicles. Energies. 2018;**11**(221):1-17

[41] Foong TM, Morganti KJ, Brear MJ, da Silva G, Yang Y, Dryer FL. The octane numbers of ethanol blended with gasoline and its surrogates. Fuel. 2014;**115**:727-739

[42] Luo M, El-Faroug MO, Yan F, Wang Y. Particulate matter and gaseous emission of hydrous ethanol gasoline blends fuel in a port injection gasoline engine. Energies. 2017;**10**:1-16

[43] https://www.arnoldclark.com/ newsroom/347-can-cars-run-on-alcohol [Accessed: 2018]

[44] Liao SY, Jiang DM, Cheng Q, Huang ZH, Wei Q. Investigation of the cold-start combustion characteristics of ethanol-gasoline blends in a constant-volume chamber. Energy & Fuels. 2005;**19**:813-819

[45] Yahuza I, Dandakouta H. A performance review of ethanol-diesel blended fuel samples in compression-ignition engine. Journal of Chemical Engineering & Process Technology. 2015;**6**(5):1-6

[46] Bossel U, Eliasson B. Energy and the Hydrogen Economy. 2003. Available from: https://afdc.energy.gov/files/ pdfs/hyd_economy_bossel_eliasson.pdf [Accessed: 2018]

[47] Azzeh SE, Marjan S, Fayaz R. Hydrogen economy and the built environment. In: Conference: World Renewable Energy Congress—Sweden, 8-13 May 2011; Linköping, Sweden. 2011

[48] Rao S, Dipak A. Review of hydrogen as a fuel in IC engines. International Journal of Science and Research. 2017;**7**:914-922

[49] Satheesh kumar C, Mohammed Shekoor T. Evaluation of emission characteristics of hydrogen as a boosting fuel in a four stroke single cylinder gasoline engine. International Journal of Engineering Research & Technology (IJERT). 2014;**3**(10):151-154

[50] Gandhi R. Use of hydrogen in internal combustion engine. International Journal of Engineering and Technical Research (IJETR). 2015;**3**(2):207-216

[51] Yousufuddin S, Mehdi SN, Masood M. Performance and combustion characteristics of a hydrogen-ethanol fuelled engine. Energy & Fuels. 2008;**22**:3355-3362

[52] Petkov T, Veziroglu TN, Sheffield JW. An outlook of hydrogen as an automotive fuel. International Journal of Hydrogen Energy. 1989;**14**:449-474

[53] Lee JT, Kim YY. The development of a dual injection hydrogen fueled engine with high power and high efficiency. In: Proceedings of the 2002 Fall Technical Conference of the ASME Internal Combustion Engine Division, ICEF2002-514, 8-11 September, New Orleans, Louisiana, USA. 2002. pp. 323-333

[54] Ma F, He Y, Deng J, Jiang L, Naeve N, Wang M, et al. Idle characteristics

of a hydrogen fueled SI engine. International Journal of Hydrogen Energy. 2011;**36**(7):4454-4460

[55] Yamin Jehan AA, Gupta HN, Bansal BB, Srivastava ON. Effect of combustion duration on the performance and emission characteristics of a spark ignition engine using hydrogen as a fuel. International Journal of Hydrogen Energy. 2000;**25**(6):581-590

[56] De Boer PCT, McLean WJ, Homan HS. Performance and emissions of hydrogen fueled internal combustion engines. International Journal of Hydrogen Energy. 1976;**1**(2):153-172

[57] Jie M, Yongkang S, Yucheng Z, Zhongil Z. Simulation and prediction on the performance of a vehicle's hydrogen engine. International Journal of Hydrogen Energy. 2003;**28**(1):77-83

[58] Ciniviz M, Köse H. Hydrogen use in internal combustion engine. International Journal of Automotive Engineering and Technologies. 2012;**1**(1):1-15

[59] Wang S, Ji C, Zhang B, Zhou X. Analysis on combustion of a hydrogen-blended gasoline engine at high loads and lean conditions. Energy Procedia. 2014;**61**:323-326

[60] D'Andrea T, Henshaw PF, KTing DS. The addition of hydrogen to a gasoline-fuelled SI engine. International Journal of Hydrogen Energy. 2004;**29**(14):1541-1552

[61] Kahraman E, Cihangir Ozcanlı S, Özerdem B. An experimental study on performance and emission characteristics of a hydrogen fuelled spark ignition engine. International Journal of Hydrogen Energy. 2007;**32**(12):2066-2072

[62] Wang S, Ji C, Zhang J, Zhang B. Improving the performance of a gasoline engine with the addition of hydrogen-oxygen mixtures.

International Journal of Hydrogen Energy. 2011;**36**(17):11164-11173

[63] Turns SR. An Introduction to Combustion Concepts and Applications. New Delhi: McGraw-Hill; 2000. 676 p

[64] Huang Z, Wang J, Liu B, Zeng K, Yu J, Jiang D. Combustion characteristics of a direct-injection engine fueled with natural gas—Hydrogen blends under different ignition timings. Fuel. 2007;**86**(3):381-387

[65] Akansu SO, Dulger A, Kahraman N, Veziroglu TN. Internal combustion engines fueled by natural gas—Hydrogen mixtures. International Journal of Hydrogen Energy. 2004;**29**(14):1527-1539

[66] Collier K, Mulligan N, Shin D, Brandon S. Emission results from the new development of a dedicated hydrogen-enriched natural gas heavy duty engine. 2005. SAE Paper No. 2005-01-0235

[67] Nanthagopal K, Subbarao R, Elango T, Baskar P, Annamalai K. Hydrogen enriched compressed natural gas (HCNG)—A futuristic fuel for internal combustion engines. Thermal Science. 2011;**15**(4):1145-1154

[68] Bell SR, Gupta M. Extension of a lean operating limit for natural gas fuelling of a spark ignition engine using hydrogen blending. Combustion Science and Technology. 1997;**123**(1-6):23-48

[69] Munshi SR, Nedelcu C, Harris J. Hydrogen blended natural gas of a operation of a heavy duty turbocharged lean burn spark ignition engine. 2004. SAE Paper No. 2004-01-2956

[70] Sandhu SS, Babu MKG, Das LM. Investigations of emission characteristics and thermal efficiency in a spark-ignition engine fuelled with natural gas—Hydrogen blends.

International Journal of Low Carbon Technologies. 2013;**8**(1):7-13

[71] Thring RH. Alternative fuels for spark-ignition engines. 1983. SAE Paper No. 831685

[72] Desoky AA, El-Emam SH. A study on the combustion of alternative fuels in spark-ignited engines. International Journal of Hydrogen Energy. 1985;**10**:497-504

[73] Cooney AP, Yeliana JJ, Worm JD, Naber JD. Combustion characterization in an internal combustion engine with ethanol gasoline blended fuels varying compression ratios and ignition timing. Energy & Fuels. 2009;**23**:2319-2324

[74] Al-Baghdadi MAS, Al-Janabi HAS. Improvement of performance and reduction of pollutant emission of a four stroke spark ignition engine fueled with hydrogen-gasoline fuel mixture. Energy Conversion and Management. 2000;**41**:77-91

[75] Zhang B, Ji C, Wang S. Performance of a hydrogen-enriched ethanol engine at unthrottled and lean conditions. Energy Conversion and Management. 2016;**114**:68-74

[76] Norman DB. Ethanol fuelda single-cylinder engine study of efficiency and exhaust emissions. 1982. SAE Paper No. 810345.1410-24

[77] Akansu SO, Tangöz S, Kahraman N, Ilhak Mİ, Açıkgöz S. Experimental study of gasoline-ethanol-hydrogen blends combustion in an SI engine. International Journal of Hydrogen Energy. 2017;**42**(40):25781-25790

[78] Shuofeng W, Changwei J, Zhang B. Effect of hydrogen addition on combustion and emissions performance of a spark-ignited ethanol engine at idle and stoichiometric conditions. International Journal of Hydrogen Energy. 2010;**35**:9205-9213

[79] Park C, Choi Y, Kim C, Oh S, Lim G, Moriyoshi Y. Performance and exhaust emission characteristics of a spark ignition engine using ethanol and ethanol-reformed gas. Fuel. 2010;**89**:2118-2125

[80] Al-Hamamre Z, Yamin J. The effect of hydrogen addition on premixed laminar acetylene-hydrogen-air and ethanol-hydrogen-air flames. International Journal of Hydrogen Energy. 2013;**38**:7499-7509

[81] Al-Baghdadi M. Hydrogen-ethanol blending as an alternative fuel for spark ignition engines. Renewable Energy. 2003;**28**:1471-1478

[82] Schefer RW. Hydrogen enrichment for improved lean flame stability. International Journal of Hydrogen Energy. 2003;**28**:1131-1141

[83] Wang J, Huang Z, Tang C, Zheng J. Effect of hydrogen addition on early flame growth of lean burn natural gas-air mixtures. International Journal of Hydrogen Energy. 2010;**35**:7246-7252

[84] Ceper B, Aydın K, Akansu SO, Kahraman N. Numerical simulation and experimental studies of a biogas fueled spark ignition engine. Energy Education Science and Technology Part A: Energy Science and Research. 2012;**28**(2):599-610

[85] Yousufuddin S, Masood M. Effect of ignition timing and compression ratio on the performance of a hydrogen-ethanol fuelled engine. International Journal of Hydrogen Energy. 2009;**34**(16):6945-6950

[86] Kumar NS, Prabhu BG, Selvan KK, Kumar RM, Kumar KM. Emission and performance characteristics of hydrogen-acetylene fuel in IC engine. International Journal of Innovative Research in Science, Engineering and Technology. 2017;**6**(3):4620-4627

[87] Tangöz S, İlhak Mİ, Akansu SO, Kahraman N. Experimental investigation of performance and emissions of an SI engine fueled by acetylene-methane and acetylene-hydrogen blends. Fresenius Environmental Bulletin. 2018;**27**:4174-4185

[88] Aziz Hairuddin A, Wandel AP, Yusaf TF. Hydrogen and natural gas comparison in diesel HCCI engines—A review. Southern Region Engineering Conference. 11-12 November 2010. Toowoomba, Australia. 2010

[89] Liu J, Wang J, Zhao H. Optimization of the injection parameters and combustion chamber geometries of a diesel/natural gas RCCI engine. Energy. 2018;**164**:837-852

[90] Jia M, Xie M, Wang T, Peng Z. The effect of injection timing and intake valve close timing on performance and emissions of diesel PCCI engine with a full engine cycle CFD simulation. Applied Energy. 2011;**88**(9):2967-2975

[91] Zheng Z, Yao M. Charge stratification to control HCCI: Experiments and CFD modeling with n-heptane as fuel. Fuel. 2009;**88**:354-365

[92] Singh AP, Agarwal AK. Low-Temperature Combustion: An Advanced Technology for Internal Combustion Engines. Singapore: Springer Nature Singapore Pte Ltd; 2018. pp. 9-41

[93] Noehre C, Andersson M, Johansson B, Hultqvist A. Characterization of partially premixed combustion. 2006. SAE Technical Paper 2006-10-16

[94] Fiveland SB, Assanis DN. A four-stroke homogeneous charge compression ignition engine simulation for combustion and performance studies. 2000. SAE Paper No. 2000-01-0332

[95] Rattanapaibule K, Aung K. Performance predictions of a hydrogen-enhanced natural gas HCCI engine. In: International Mechanical Engineering Congress and Exposition (IMECE2005), Florida. 2005. pp. 289-294

[96] Caton J. Thermodynamic advantages of low temperature combustion (LTC) engines using low heat rejection (LHR) concepts. 2011. SAE Technical Paper 2011-01-0312

[97] Asad U, Divekar P, Zheng M, Tjong J. Low temperature combustion strategies for compression ignition engines: Operability limits and challenges. 2013. SAE Technical Paper. 2013-01-0283

[98] Okude K, Mori K, Shiino S, Moriya T. Premixed compression ignition (PCI) combustion for simultaneous reduction of NOx and soot in diesel engine. 2004. SAE Technical Paper Series. 2004-01-1907

[99] Fiveland SB, Agama R, Christensen M, Johansson B, Hiltner L, Maus F, et al. Experimental and simulated results detailing the sensitivity of natural gas hcci engines to fuel composition. 2001. SAE Technical Paper 2001-01-3609

[100] Jamsran N, Putrasari N, Lim O. A computational study on the autoignition characteristics of an HCCI engine fueled with natural gas. Journal of Natural Gas Science and Engineering. 2016;**29**:469-478

[101] Morsy MH. Ignition control of methane fueled homogeneous charge compression ignition engines using additives. Fuel. 2007;**86**:533-540

[102] Yousefzadeh A, Jahanian O. Using detailed chemical kinetics 3D-CFD model to investigate combustion phase of a CNG-HCCI engine according to control strategy requirements. Energy Conversion and Management. 2017;**133**:524-534

[103] Zheng J, Caton JA. Effects of operating parameters on nitrogen oxides emissions for a natural gas fueled homogeneous charged compression ignition engine (HCCI): Results from a thermodynamic model with detailed chemistry. Applied Energy. 2012;**92**:386-394

[104] Christensen M, Johansson B, Einewall P. Homogeneous charge compression ignition (HCCI) using isooctane, ethanol and natural gas: A comparison with spark ignition operation. 1997. SAE Technical Paper 972874

[105] Maurya RK, Agarwal AK. Experimental investigations of performance, combustion and emission characteristics of ethanol and methanol fueled HCCI engine. Fuel Processing Technology. 2014;**126**:30-48

[106] Bahri B, Aziz AA, Shahbakhti M, Said MFM. Understanding and detecting misfire in an HCCI engine fuelled with ethanol. Applied Energy. 2013;**108**:24-33

[107] Maurya RK, Agarwal AK. Experimental study of combustion and emission characteristics of ethanol fuelled port injected homogeneous charge compression ignition (HCCI) combustion engine. Applied Energy. 2012;**88**:1169-1180

[108] Viggiano A, Magi V. A comprehensive investigation on the emissions of ethanol HCCI engines. Applied Energy. 2012;**93**:277-287

[109] Sudheesh K, Mallikarjuna JM. Development of an exhaust gas recirculation strategy for an acetylene-fuelled homogeneous charge compression ignition engine. Proceedings of the Institution of Mechanical Engineers, Part D: Journal of Automobile Engineering. 2010;**224**:941-952

[110] Puranam S, Steeper R. The effect of acetylene on iso-octane combustion in an HCCI engine with NVO. SAE International Journal of Engines. 2012;**5**(4):1551-1560

[111] Nathan SS, Mallikarjuna J, Ramesh A. HCCI engine operation with acetylene the fuel. 2008. SAE Technical Paper 2008-28-0032

[112] Nathan SS, Mallikarjuna JM, Ramesh A. Effects of charge temperature and exhaust gas re-circulation on combustion and emission characteristics of an acetylene fuelled HCCI engine. Fuel. 2010;**89**:515-521

[113] Aithal SM. Prediction of voltage signature in a homogeneous charge compression ignition (HCCI) engine fueled with propane and acetylene. Combustion Science and Technology. 2013;**185**:1184-1201

[114] Sudheesh K, Mallikarjuna JM. Diethyl ether as an ignition improver for acetylene-fuelled homogeneous charge compression ignition operation: An experimental investigation. International Journal of Sustainable Energy. 2015;**34**(9):561-577

[115] Guo H, Hosseini V, Neill WS, Chippior WL, Dumitrescu CE. An experimental study on the effect of hydrogen enrichment on diesel fueled HCCI combustion. International Journal of Hydrogen Energy. 2011;**36**(21):13820-13830

[116] Guo H, Neill WS. The effect of hydrogen addition on combustion and emission characteristics of an n-heptane fuelled HCCI engine. International Journal of Hydrogen Energy. 2013;**38**(26):11429-11437

[117] Antunes JMG, Mikalsen R, Roskilly AP. An investigation of hydrogen-fuelled HCCI engine performance and operation. International

Journal of Hydrogen Energy. 2008;**33**(20):5823-5828

[118] Ibrahim MM, Ramesh AA. Investigations on the effects of intake temperature and charge dilution in a hydrogen fueled HCCI engine. International Journal of Hydrogen Energy. 2014;**39**(26):14097-14108

[119] Gowda BD, Echekki T. Complex injection strategies for hydrogen-fueled HCCI engines. Fuel. 2012;**97**:418-427

[120] Maurya RK, Saxena MR. Characterization of ringing intensity in a hydrogen-fueled HCCI engine. International Journal of Hydrogen Energy. 2018;**43**(19):9423-9437

[121] Shudo T, Yamada H. Hydrogen as an ignition-controlling agent for HCCI combustion engine by suppressing the low-temperature oxidation. International Journal of Hydrogen Energy. 2007;**32**(14):3066-3072

[122] Kozlov VE, Chechet IV, Matveev AG, Titova NS, Starik AM. Modeling study of combustion and pollutant formation in HCCI engine operating on hydrogen rich fuel blends. International Journal of Hydrogen Energy. 2016;**41**(5):3689-3700

[123] Poorghasemi K, Saray RK, Ansari E, Irdmousa BK, Shahbakhti M, Naber ND. Effect of diesel injection strategies on natural gas/diesel RCCI combustion characteristics in a light duty diesel engine. Applied Energy. 2017;**199**:430-446

[124] Kakaee AH, Rahnama P, Paykani A. Influence of fuel composition on combustion and emissions characteristics of natural gas/diesel RCCI engine. Journal of Natural Gas Science and Engineering. 2015;**25**:58-65

[125] Gharehghani A, Hosseini R, Mirsalim M, Jazayeri A, Yusaf T. An experimental study on reactivity

controlled compression ignition engine fueled with biodiesel/natural gas. Energy. 2015;**89**:558-567

[126] Ansari E, Shahbakhti M, Naber J. Optimization of performance and operational cost for a dual mode diesel-natural gas RCCI and diesel combustion engine. Applied Energy. 2018;**231**:549-561

[127] Walker NR, Wissink ML, DelVescovo DA, Reitz RD. Natural gas for high load dual-fuel reactivity controlled compression ignition (RCCI) in heavy-duty engines. In: Proceedings of the ASME Internal Combustion Engine Division, Fall Technical Conference; Vol. 1. ASME, V001T03A016. 2014

[128] Olmeda P, García A, Monsalve-Serrano J, Sari RL. Experimental investigation on RCCI heat transfer in a light-duty diesel engine with different fuels: Comparison versus conventional diesel combustion. Applied Thermal Engineering. 2018;**144**:424-436

[129] Khatamnejad H, Khalilarya SH, Jafarmadar S, Mirsalim M. The effect of high-reactivity fuel injection parameters on combustion features and exhaust emission characteristics in a natural gas–diesel RCCI engine at part load condition. International Journal of Green Energy. 2018;**15**(13):874-888

[130] Jia Z, Denbratt I. Experimental investigation of natural gas-diesel dual-fuel RCCI in a heavy-duty engine. SAE International Journal of Engines. 2015;**8**(2):797-807

[131] Shim E, Park H, Bae C. Intake air strategy for low HC and CO emissions in dual-fuel (CNG-diesel) premixed charge compression ignition engine. Applied Energy. 2018;**225**:1068-1077

[132] Park H, Shim E, Bae C. Improvement of combustion and emissions with exhaust gas recirculation

in a natural gas-diesel dual-fuel premixed charge compression ignition engine at low load operations. Fuel. 2019;**235**:763-774

[133] Esfahanian V, Salahi MM, Gharehghani A, Mirsalim M. Extending the lean operating range of a premixed charged compression ignition natural gas engine using a pre-chamber. Energy. 2017;**119**:1181-1194

[134] Dempsey AB, Das Adhikary B, Viswanathan S, Reitz RD. Reactivity controlled compression ignition (RCCI) using premixed hydrated ethanol and direct injection diesel. In: Proceedings of the ASME Internal Combustion Engine Division Fall Technical Conference (ICEF). 2011. pp. 963-975

[135] Qian Y, Wang X, Zhu L, Lu X. Experimental studies on combustion and emissions of RCCI (reactivity controlled compression ignition) with gasoline/n-heptane and ethanol/n-heptane as fuels. Energy. 2015;**88**:584-594

[136] Loaiza JCV, Sanchez FZ, Braga SL. Combustion study of reactivity-controlled compression ignition (RCCI) for the mixture of diesel fuel and ethanol in a rapid compression machine. Journal of the Brazilian Society of Mechanical Sciences and Engineering. 2016;**38**(4):1073-1085

[137] Liu H, Ma G, Hu B, Zheng Z, Yao M. Effects of port injection of hydrous ethanol on combustion and emission characteristics in dual-fuel reactivity controlled compression ignition (RCCI) mode. Energy. 2016;**145**:592-602

[138] Park SH, Shin D, Park J. Effect of ethanol fraction on the combustion and emission characteristics of a dimethyl ether-ethanol dual-fuel reactivity controlled compression ignition engine. Applied Energy. 2016;**182**:243-252

[139] Elzahaby AM, Elkelawy M, Bastawissi HAE, El-Malla SM, Naceb AMM. Kinetic modeling and experimental study on the combustion, performance and emission characteristics of a PCCI engine fueled with ethanol-diesel blends. Egyptian Journal of Petroleum. 2018;**27**(4):927-937

[140] Natarajan S, Shankar SA, Sundareswaran M. Early injected PCCI engine fuelled with bio ethanol and diesel blends: An experimental investigation. Energy Procedia. 2017;**105**:358-366

[141] Saravanan S, Pitchandi K, Suresh G. An experimental study on premixed charge compression ignition-direct ignition engine fueled with ethanol and gasohol. Alexandria Engineering Journal. 2015;**54**(4):897-904

[142] Mancaruso E, Vaglieco BM. Spectroscopic analysis of the phases of premixed combustion in a compression ignition engine fuelled with diesel and ethanol. Applied Energy. 2015;**143**:164-175

[143] Kokjohn S, Splitter DA, Reitz RD, Manente V, Johansson B. Modeling charge preparation and combustion in diesel fuel, ethanol, and dual-fuel PCCI engines. Atomization and Sprays. 2011;**21**:107-119

[144] Noh HK, No SY. Effect of bioethanol on combustion and emissions in advanced CI engines: HCCI, PPC and GCI mode: A review. Applied Energy. 2017;**208**:782-802

[145] Sartbaeva A, Kuznetsov VL, Wells S, Edwards P. Hydrogen nexus in a sustainable energy future. Energy & Environmental Science. 2008;**1**(1):79-85

[146] Lovel WG. Knocking characteristics of hydrocarbons. Journal of Industrial and Engineering Chemistry. 1948;**40**(12):2388-2438

[147] Hoseinpour M, Sadrnia H, Tabasizadeh M, Ghobadian B. Energy

and exergy analyses of a diesel engine fueled with diesel, biodiesel-diesel blend and gasoline fumigation. Energy. 2017;**141**:2408-2420

[148] Ji C, Shi L, Wang S, Cong X, Su T, Yu M. Investigation on performance of a spark-ignition engine fueled with dimethyl ether and gasoline mixtures under idle and stoichiometric conditions. Energy. 2017;**126**:335-342

[149] Özcan H. Hydrogen enrichment effects on the second law analysis of a lean burn natural gas engine. International Journal of Hydrogen Energy. 2010;**35**(3):1443-1452

[150] Papagiannakis RG, Rakopoulos CD, Hountalas DT, Rakopoulos DC. Emission characteristics of high speed, dual fuel, compression ignition engine operating in a wide range of natural gas/diesel fuel proportions. 7th International Symposium on Alcohol Fuels. 2010;**89**(7):1397-1406

[151] Greenwood JB, Erickson PA, Hwang J, Jordan EA. Experimental results of hydrogen enrichment of ethanol in an ultra-lean internal combustion engine. International Journal of Hydrogen Energy. 2014;**39**:12980-12990

[152] Catapano F, Di Iorio S, Magno A, Sementa P, Vaglieco BM. A comprehensive analysis of the effect of ethanol, methane and methane-hydrogen blend on the combustion process in a PFI (port fuel injection) engine. Energy. 2015;**88**:101-110

[153] Pulkrabek WW. Engineering Fundamentals of the Internal Combustion Engine. 2nd ed. New Jersey: Prentice Hall; 2003

[154] Edwards R, Larivé JF, Rickeard D, Weindorf W. Well-to-Wheels analysis of future automotive fuels and powertrains in the European context well-to-tank (WTT) report version 4.a,

Joint Research Centre of the European Commission, Luxembourg: Publications Office of the European Union, 2014

Chapter 4

A New Lightweight Material for Possible Engine Parts Manufacture

Akaehomen O. Akii Ibhadode and Raphael S. Ebhojiaye

Abstract

In the current drive for cleaner energy use, the application of lightweight materials in internal combustion engines becomes imperative as it makes for greater fuel efficiency which results in pollution reduction. This chapter reviews the materials being developed in this direction and then discusses a particularly new lightweight hybrid composite material made of palm kernel shell (PKS) and periwinkle shell (PS) particles as reinforcements in commercially pure aluminium matrix. The fabricated composite had significantly improved properties over the commercially pure aluminium and was used to produce a lightweight engine block. Preliminary performance test results show that the hybrid aluminium composite may be suitable for some engine parts manufacture such as an engine block. Weight analysis carried out on an existing engine shows that the use of this new material in the manufacture of the engine block, cylinder head, piston and connecting rod could give a potential weight reduction of over 25% when used in place of conventional materials. Also, the results show that potential energy cost saving of over 62% could be achieved when this new material is used. However, further work is needed to properly ascertain its areas of specific application.

Keywords: lightweighting, aluminium composite, palm kernel shell, periwinkle shell, engine block

1. Introduction

1.1 The internal combustion engine and the environment

An overwhelming majority of transport vehicles are driven by internal combustion (IC) engines which use fossil fuels as the propellant. Fossil fuel used in vehicles is known to be one cause of environmental degradation. The global need to stem environmental degradation, climaxed at the 2015 United Nations Climate Change Conference, COP 21/CMP 11 held in Paris, France. The Conference came up with the target of limiting global warming to below 2°C. This tells us that additional and more stringent measures are likely to be imposed on fossil fuels in the future. This will negatively impact on the use of internal combustion (IC) engines in vehicles. Even now, new targets on fuel economy are being set: 'by 2021, new cars in Europe should emit no more than 95 g/km of CO_2, representing a reduction of 40% compared with the fleet average of 158.7 g/km in 2007. At the same time, CAFÉ regulations in the US will require light vehicles to achieve 54.5 mpg (23.2 km/l) by 2025. These developments make manufacturing costs more important than ever. With approximately 87 million vehicles sold in 2015 and this number expected to rise to

115 million by 2030, introducing lightweight materials and innovative processes will enable car manufacturers to meet these ambitious targets' [1].

One method used by the automobile industry to tackle this challenge posed by new emission regulations is for IC engines to burn fuel more efficiently. Lightweighting, which is building lighter vehicles, is used to have better fuel efficiency [2].

1.2 Purpose of lightweighting

The main purpose of lightweighting vehicles is to increase fuel efficiency [3]. For example, while a conventional gasoline car weighing about 750 kg may have a fuel economy of 22 km/l, a Shell eco-marathon [4] urban-concept gasoline vehicle weighing about 45 kg may have a fuel efficiency of over 400 km/l. This vividly shows the benefit of weight reduction on fuel economy.

Other reasons for lightweighting include:

i. To achieve better vehicle performance to give better acceleration, braking and handling.

ii. To have greater load-carrying capacity to engine power ratio.

1.3 Methods of lightweighting

The methods of automobile lightweighting include:

i. The use of lighter materials such as high-strength steel, aluminium, magnesium, lightweight composites and plastics [5].

ii. The use of optimum structural designs such as boxlike geometries to achieve greater rigidity to weight ratio. **Figure 1** shows a car bumper with stacked-joined hexagonal lightweight pipes [6].

iii. The use of hot-forming techniques including hot-stamping to produce ultra-high strength parts such as used for reinforcements for vehicle doors, bumper beams, etc., which can reduce weight to the tune of 50% compared to cold-formed parts [1].

iv. 'Body-in-white' (BIW) technique in which the unpainted metal components are welded together to form the vehicle's body, which can account for approximately 50% of weight saving [1].

Figure 1.
The use of stacked-joined hexagonal lightweight pipes for a car bumper [6].

1.4 Lightweighting materials

Table 1 shows the major types of lightweighting materials used in the automotive industry [7]. It shows that weight reduction could be as high as 70% when carbon fibre composites and magnesium are used to replace mild steel.

1.5 Fabricated metal composites

Lightweight engineering materials require high strength, long life, high wear and corrosion resistance. In the current drive for cleaner energy use, the application of lightweight materials in internal combustion engines becomes imperative as it makes for greater fuel efficiency which results in pollution reduction. Unfortunately, there are no naturally occurring engineering materials that perfectly possess all these properties at the same time. To achieve some of these qualities in a single engineering material, dissimilar materials with unique characteristics can be made together as alloys or composite materials to produce improved performance characteristics. 'Such fabricated composite materials find very wide application [8] in electronics, sporting goods, aerospace parts, consumer goods, marine and oil industries, automobile components like IC engine parts, etc. because of the high requirements in product performance and rise in global market demand of lightweight components' [9].

1.6 A new material for IC engine parts

Material selection for IC engine components has been one active area of research. Proper selection of materials for the production of IC engines is critical because of the high temperature and fatigue stresses the engine is subjected to. In selecting a suitable material for IC engine components, important parameters or factors such as the physical properties (i.e. melting temperature, hardness, creep and flow, corrosion, weight and vibration absorption) and mechanical properties (i.e. strength, strain, elastic modulus, ultimate tensile strength, yield strength, elongation, fatigue strength, pressure tightness and machinability) as well as availability and cost of the materials must be considered.

Palm kernel and periwinkle are common sources of foods and are harvested and processed in very large quantities in most parts of Southern Nigeria. However, their shells are disposed of indiscriminately giving rise to environmental challenges to communities where they are processed because palm kernel shell (PKS) and periwinkle shell (PS) do readily undergo biological degradation [10, 11].

Lightweighting material	Mass reduction with respect to mild steel
Carbon fibre composites	50–70%
Magnesium	30–70%
Aluminium and its matrix composites	30–60%
Titanium	40–55%
Glass fibre composites	25–35%
Advanced high-strength steel	15–25%
High-strength steel	10–28%

Table 1.
Comparison of different lightweighting materials (Adapted from [7]).

Aluminium and magnesium alloys are commonly used as lightweight engineering materials, and they are gradually taking over from cast iron and steel in the automobile and aerospace industries and defence applications [12, 13]. To improve the properties of aluminium, they are usually alloyed with other very expensive elements like chromium, nickel, molybdenum, silicon, boron, vanadium, etc. However, these alloying metallic elements are expensive and difficult to source in most parts of the developing world such as Nigeria. This study is directed at exploring alternative substitute materials for production of metal matrix composites (MMC) that can be suitable for the production of automobile parts [14]. Thus, the purpose of the study is to develop a hybrid composite material with PKS and PS as the reinforcement particles in aluminium matrix, for the production of appropriate IC engine parts. In this work, the engine block is used as case study because it is particularly important in that it houses the power-producing chamber and forms the fixed point for all other members of the engine. It is hoped that this could be a cheaper alternative to existing aluminium alloy engine parts.

2. Methodology

2.1 Materials

The following materials were used:

2.1.1 Commercially pure aluminium ingot

Commercially pure aluminium pieces cut from an ingot with the composition shown in **Table 2** was used.

2.1.2 Palm kernel shell (PKS)

Palm kernel shell, a common agro waste (**Figure 2(a)**), was collected from a palm oil processing factory in Uselu Community in Benin City, Nigeria. The PKS was washed thoroughly with detergent and spread in the sun to dry. The dry shell was pulverised into powder of particle size 425 μm. The PKS powder is shown in **Figure 2(b)**. PKS is reported to have chemical properties which include 8.786% calcium oxide (CaO), 6.254% potassium oxide (K_2O), 54.81% silicon oxide (SiO_2), 6.108% magnesium oxide (MgO), 0.362% iron oxide, 11.40% alumina (Al_2O_3), 33% charcoal, 45% pyroligneous liquor and 22% combustible materials [15].

2.1.3 Periwinkle shell (PS)

The periwinkle shell (**Figure 3(a)**), another agro waste belonging to the pozzolanic group, is available in large quantities in the estuaries and mangrove swamp forest of the South-South region of Nigeria [16]. A large quantity of PS was sourced

Mg	Si	Mn	Cu	Zn	Ti
0.00118	<0.03456	0.00058	<0.00035	0.00058	0.00438
Fe	Na	B	Sn	Pb	Al%
<0.09775	0.00118	0.00024	0.01160	0.00116	99.85

Table 2.
Chemical composition of the commercially pure aluminium.

Figure 2.
Palm kernel shell (PKS). (a) Broken shells and (b) 425 μm powder.

Figure 3.
Periwinkle shell. (a) Shells and (b) 75 μm powder.

from its dump site around New Benin Market in Benin City, Nigeria. The PS was soaked in a mixture of detergent and water for 24 h and later washed and spread in the sun to dry. The clean PS was then pulverised with an electrically driven crushing machine into a powder of particle size 75 μm as shown in **Figure 3(b)**. PS is reported to have a chemical composition of calcium content of 8.25×10^3 mg/100 g, potassium content of 5.50 mg/100 g, sodium content of 5.30 mg/100 g and phosphorus content of 2.55 mg/100 g [16].

The hybrid aluminium composite material used in this study was made up of appropriate percentages of commercially pure aluminium and particular percentages of PKS and PS [14, 17].

2.2 Methods

The hybrid composite material of aluminium metal matrix using agro wastes of palm kernel shell (PKS) and periwinkle shell (PS) powder as the reinforcement agents was designed using the central composite design (CCD) of the response surface methodology (RSM). In order to aid homogenous mix of the reinforcement materials in the matrix, the hybrid composite was formulated and fabricated by stir casting [18, 19] process using a stir casting machine.

2.2.1 Design of the experiment

The experimental design for the study was done to optimise operating conditions (i.e. factor levels of the input variables) of the fabricated composite with respect to the predicted response parameters. The CCD method was selected because it allows the addition of axial runs which in turn allows the quadratic terms to be incorporated into the model. Three input variables were used in the design: pure aluminium ingot, PKS and PS. A total of six response variables were investigated: wear rate (g/s), creep rate (% elongation/h), density (kg/m³), tensile

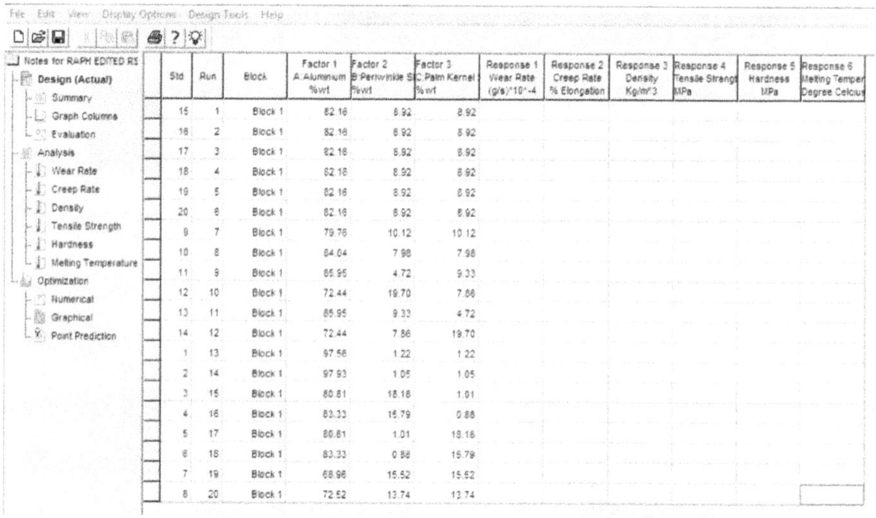

File Edit View Display Options Design Tools Help

Std	Run	Block	Factor 1 A:Aluminum %wt	Factor 2 B:Periwinkle S %wt	Factor 3 C:Palm Kernel %wt	Response 1 Wear Rate (g/s)*10^-4	Response 2 Creep Rate % Elongation	Response 3 Density Kg/m^3	Response 4 Tensile Strength MPa	Response 5 Hardness MPa	Response 6 Melting Temper Degree Celcius
15	1	Block 1	82.16	8.92	8.92						
16	2	Block 1	82.16	8.92	8.92						
17	3	Block 1	82.16	8.92	8.92						
18	4	Block 1	82.16	8.92	8.92						
19	5	Block 1	82.16	8.92	8.92						
20	6	Block 1	82.16	8.92	8.92						
9	7	Block 1	79.76	10.12	10.12						
10	8	Block 1	84.04	7.98	7.98						
11	9	Block 1	85.95	4.72	9.33						
12	10	Block 1	72.44	19.70	7.86						
13	11	Block 1	85.95	9.33	4.72						
14	12	Block 1	72.44	7.86	19.70						
1	13	Block 1	97.56	1.22	1.22						
2	14	Block 1	97.93	1.05	1.05						
3	15	Block 1	80.81	18.18	1.01						
4	16	Block 1	83.33	15.79	0.88						
5	17	Block 1	80.81	1.01	18.18						
6	18	Block 1	83.33	0.88	15.79						
7	19	Block 1	68.96	15.52	15.52						
8	20	Block 1	72.52	13.74	13.74						

Notes for RAPH EDITED RS
Design (Actual)
 Summary
 Graph Columns
 Evaluation
Analysis
 Wear Rate
 Creep Rate
 Density
 Tensile Strength
 Hardness
 Melting Temperature
Optimization
 Numerical
 Graphical
 Point Prediction

Figure 4.
Computer interface of the preliminary design of experiment.

strength (MPa), hardness (BHN) and melting temperature (°C). **Figure 4** shows the computer interface of the preliminary experimental design using the CCD.

2.2.2 Fabrication of the aluminium composite material

The commercially pure aluminium ingot was cut into smaller sizes. The smaller pieces of the ingots were soaked with detergent and hot water at 50°C for 10 h and then washed thoroughly with a hard brush and later rinsed in clean water. The ingots were dried at about 70°C for 1.5 hrs. A digital electronic scale with an accuracy of 0.01 g was used to weigh the ingots into the various percentage weights as obtained from the design of experiment using CCD of RSM.

The crucible furnace was initially preheated to 100°C. The pure aluminium ingots that were measured for the different experimental specimen weights were preheated for about 1 h at a temperature of 450°C [20]. The PS and PKS reinforcement particles were also preheated to 250°C to remove moisture and oil contents [21]. For the first sample labelled A-1, aluminium ingot of 87.5 wt.% ~262.5 g was charged into the graphite crucible pot with 1 wt.% of magnesium powder [22, 23] as the wetting agent. The graphite crucible was then placed inside the crucible furnace and heated to a temperature of 750°C [13, 24, 25]. At that temperature the solid aluminium ingots had melted into molten form. The melt was allowed to cool in the furnace to a slurry form (semisolid state) at a temperature of about 600°C [25]. At that slurry temperature of the molten metal, the preheated PKS and PS particles were added and stirred manually for about 7 min with the stir casting machine. The PKS particles were charred into particles of black carbon due to the high temperature of 600°C at which the PKS particles were introduced into the molten matrix. The resultant composite mixture was further heated from the slurry state to the molten state at about 720°C [25] and mechanically stirred with the stir casting machine at 400 rpm for about 10 min to form a fine vortex [26]. The molten composite mixture was then poured into prepared permanent moulds made of mild steel to form the cast composite specimen. This process was repeated thrice for all the responses. Pure aluminium ingots without the addition of reinforcement materials were charged into the crucible pot to produce a molten form that was cast in the die mould. This was used as the specimen for the control of experiment. The specimens produced including the control specimens were tested for the six responses, and the values were recorded. The values obtained from the laboratory tests for the specimens were modelled and optimised to obtain the blend ratio using RSM.

2.3 Practical application of the developed composite material

2.3.1 Production of the engine block

The developed composite material was used as engineering material to produce the engine block of a bush cutting machine shown in **Figure 5** in order to conduct practical application assessment on the material to ascertain its applicability in real-life situation. Performance test was carried out on the produced engine block.

The engine block was produced by casting and thereafter machined. The material was initially cast into a box dimension of 80 mm × 80 mm × 110 mm by the stir casting process. Thereafter, the cast material was machined using the centre lathe and milling machine. **Figure 6** shows the dimensions of the produced engine block.

Figure 5.
The bought-out bush cutting machine.

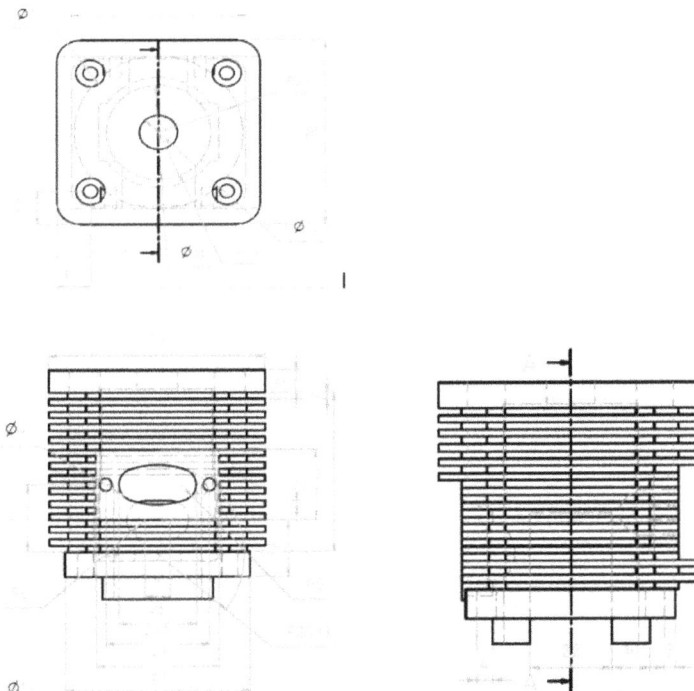

Figure 6.
The engine block of the bush cutting machine.

Figure 7.
Produced engine block in bush cutting machine.

2.3.2 Performance test of the produced engine block

The fabricated engine block and cylinder liner were assembled into the engine as shown in **Figure 7**. It was thereafter tested for 1 h. The temperature of the engine block was taken at 1 min intervals. This was repeated for the control cylinder block and liner. The bought-out bush cutting engine block was used as the control engine block.

2.3.3 Lightweighting and cost saving analyses

Lightweighting and energy cost saving analyses were carried out on an existing gasoline engine used for powering a newly designed lightweight utility vehicle. The engine is rated at 7.1 kW at 5500 rpm with a maximum torque of 18.0 Nm at 3500 rpm. The new material was hypothetically used to replace the conventional materials used for the engine block, cylinder head, piston and connecting rod. The potential weight reduction of the engine was then determined. The potential cost saving in energy was also determined.

3. Results and discussion

3.1 Optimal composition

3.1.1 Design of experiments

Ebhojiaye et al. [17] developed the hybrid composite material of palm kernel shell (PKS) and periwinkle shell (PS) particles as reinforcements in pure aluminium matrix. The central composite design (CCD) of the response surface methodology (RSM) was used to carry out the design of experiment (DoE) as shown in **Figure 4**. Stir casting method was used to fabricate the specimens. The design of experiments by the CCD gave 20 runs (experimental samples) as shown in **Table 3**. The runs were replicated three times each, bringing the total number of runs to 60 for each

Std	Run	Type	Factor 1 (aluminium, % wt)	Factor 2 (periwinkle shell, % wt)	Factor 2 (palm kernel shell, % wt)	Response 1 (wear rate (g/s) 10^{-4})	Response 2 (creep rate, % elongation/h)	Response 3 (density, kg/m^3)	Response 4 (tensile strength, MPa)	Response 5 (hardness BHN)	Response 6 (melting temperature, °C)
15	1	Centre	82.16	8.92	8.92	A1	A2	A3	A4	A5	A6
16	2	Centre	82.16	8.92	8.92	B1	B2	B3	B4	B5	B6
17	3	Centre	82.16	8.92	8.92	C1	C2	C3	C4	C5	C6
18	4	Centre	82.16	8.92	8.92	D1	D2	D3	D4	D5	D6
19	5	Centre	82.16	8.92	8.92	E1	E2	E3	E4	E5	E6
20	6	Centre	82.16	8.92	8.92	F1	F2	F3	F4	F5	F6
9	7	Axial	79.76	10.12	10.12	G1	G2	G3	G4	G5	G6
10	8	Axial	84.04	7.98	7.98	H1	H2	H3	H4	H5	H6
11	9	Axial	85.95	-4.72	9.33	I1	I2	I3	I4	I5	I6
12	10	Axial	72.44	19.70	7.86	J1	J2	J3	J4	J5	J6
13	11	Axial	85.95	9.33	-4.72	K1	K2	K3	K4	K5	K6
14	12	Axial	72.44	7.86	19.70	L1	L2	L3	L4	L5	L6
1	13	Fact	97.56	1.22	1.22	M1	M2	M3	M4	M5	M6
2	14	Fact	97.93	1.05	1.05	N1	N2	N3	N4	N5	N6
3	15	Fact	80.81	18.18	1.01	O1	O2	O3	O4	O5	O6
4	16	Fact	83.33	15.79	0.88	P1	P2	P3	P4	P5	P6
5	17	Fact	80.81	1.01	18.18	Q1	Q2	Q3	Q4	Q5	Q6
6	18	Fact	83.33	0.88	15.79	R1	R2	R3	R4	R5	R6
7	19	Fact	68.96	15.52	15.52	S1	S2	S3	S4	S5	S6
8	20	Fact	72.52	13.74	13.74	T1	T2	T3	T4	T5	T6

Table 3.
Assigned letter codes to the response parameters.

of the six responses considered, and 360 specimens were fabricated in all. Three experimental values were obtained for each of the 120 runs for the wear rate, creep rate, density, tensile strength, hardness and melting temperature. The average values were determined and recorded. Control specimens with 100 wt.% pure aluminium matrix, 0 wt.% of PKS and PS reinforcement particles were prepared. The results showed that the reinforcement particles had significant improvement on mechanical properties of the pure aluminium as shown in **Table 4** for hardness test of the various 20 fabricated composites and that of the control commercially pure aluminium.

3.1.2 RSM analysis

In optimising the results obtained for the six responses using the central composite design (CCD) of the response surface methodology (RSM), three of the responses [i.e. wear rate (y_1), creep rate (y_2) and density (y_3)] were minimised, while the other three responses [i.e. tensile strength (y_4), hardness (y_5) and melting temperature (y_6)] were maximised. The optimal equations obtained for the actual factors [commercially pure aluminium (A), periwinkle shell (B) and palm kernel shell (C)] for the six responses are shown in Eqs. (1)–(6).

1. Wear rate $(y\ 1)$

$$y_1 = 25.021 - 0.503A - 0.315B - 0.278C + 0.00273AB + 0.00205AC \\ + 0.00279BC + 0.00268\,A^2 + 0.00265\,B^2 + 0.00343\,C^2 \tag{1}$$

2. Creep rate $(y\ 2)$

$$y_2 = 2303.565 - 49.151A - 14.268B - 9.448C + 0.137AB \\ + 0.108AC - 0.039BC + 0.263\,A^2 + 0.179\,B^2 + 0.042\,C^2 \tag{2}$$

3. Density $(y\ 3)$

$$y_3 = -4283.948 + 152.915A + 50.465B - 86.331C - 0.607AB \\ + 0.925AC + 0.253BC - 0.839\,A^2 + 0.252\,B^2 + 0.035\,C^2 \tag{3}$$

4. Tensile strength $(y\ 4)$

$$y_4 = -198.881 + 6.437A + 7.726B - 6.745C - 0.081AB + 0.088AC \\ + 0.032BC - 0.036\,A^2 - 0545\,B^2 + 0.066\,C^2 \tag{4}$$

5. Hardness $(y\ 5)$

$$y_5 = -115.362 - 0.774A + 6.128B + 29.139C - 0.180AB - 0.079AC \\ - 1.159BC + 0.032\,A^2 - 0.548\,B^2 - 0.479\,C^2 \tag{5}$$

6. Melting temperature $(y\ 6)$

$$y_6 = 475.599 + 9.379A - 4.784B + 17.014C + 0.119AB - 0.096AC \\ - 0.109BC - 0.058\,A^2 - 0.219\,B^2 + 0.302\,C^2 \tag{6}$$

The result of the RSM analysis gave an optimal blend ratio of 80.98 wt.% aluminium, 13.55 wt.% periwinkle shell and 5.47 wt.% palm kernel shell particles. This optimum blend ratio when used to produce an engineering composite material

Composite label	Diameter of indenter (D) mm	Diameter of indentation (d) mm	Load (Kg)	BHN (HBS 10/1000)
A5	10.00	2.08	1000	291.08
B5	10.00	2.12	1000	280.07
C5	10.00	2.11	1000	282.77
D5	10.00	2.10	1000	285.49
E5	10.00	2.08	1000	291.08
F5	10.00	2.11	1000	282.77
G5	10.00	3.20	1000	121.07
H5	10.00	1.75	1000	412.54
I5	10.00	2.54	1000	194.12
J5	10.00	3.00	1000	138.00
K5	10.00	2.77	1000	162.69
L5	10.00	2.50	1000	200.48
M5	10.00	3.00	1000	138.00
N5	10.00	3.40	1000	106.86
O5	10.00	1.98	1000	321.56
P5	10.00	1.97	1000	342.86
Q5	10.00	2.05	1000	299.75
R5	10.00	2.13	1000	277.42
S5	10.00	2.05	1000	299.75
T5	10.00	2.50	1000	200.48
U5	10.00	2.78	1000	98.50

Table 4.
Results of hardness test of the fabricated composites and pure aluminium.

will have the following properties: wear rate of 0.62×10^{-4} g/s, creep rate of 19.09% elongation/h, density of 2598.62 kg/m^3, tensile strength of 94.04 MPa, hardness of 278.83BHN and melting temperature of 935°C. This solution that was selected with the aid of the design expert software 7.01 has a desirability value of 97.3%.

3.2 Properties of the fabricated aluminium composite

3.2.1 Chemical compositions

Table 5 shows the comparison of the chemical compositions of the as-received commercially pure aluminium with the fabricated aluminium composite. The as-received composition was given by the manufacturer, while that of the fabricated composite was determined in our laboratory using the EPA Method 3050B (concentrated HCl, HNO$_3$ and HClO$_4$). The results show that all the elements detected in the as-received sample increased in the fabricated composite (except for the ones that were not detected: boron and tin), while the aluminium decreased in the fabricated composite. However, the elements, cadmium, calcium, chromium and potassium, were added into the fabricated composite. The increase of elements and additions could have come from the palm kernel shell (PKS) and periwinkle shell (PS) as seen from Sections 2.1.2 and 2.1.3, especially for the elements silicon, calcium, potassium, iron and sodium.

Element	As-received	Fabricated aluminium composite	Remark
Magnesium (Mg)	0.00118	0.027	Increased
Silicon (Si)	<0.03456	0.28	Increased
Manganese (Mn)	0.00058	0.19	Increased
Copper (Cu)	<0.00035	0.02	Increased
Zinc (Zn)	0.00058	0.03	Increased
Titanium (Ti)	0.00438	—	Not detected
Iron (Fe)	0.09775	0.26	Increased
Sodium (Na)	0.00118	0.19	Increased
Boron (B)	0.00024	—	Not detected
Tin (Sn)	0.0116	—	Not detected
Lead (Pb)	0.00116	0.06	Increased
Aluminium (Al)	99.85	96.90	Decreased
Cadmium (Cd)	—	0.001	Added
Calcium (Ca)	—	0.11	Added
Chromium (Cr)	—	0.0001	Added
Potassium (K)	—	0.16	Added

Table 5.
Chemical compositions of the as-received and fabricated composite samples.

3.2.2 Physical and mechanical properties

The optimum blend ratio obtained from the RSM analysis, when used to produce an engineering composite material, gave the properties as shown in **Table 6**.
This is compared with properties of the control commercially pure aluminium
tested under same conditions and those predicted by the RSM analysis. The table
shows that there is improvement on properties of the fabricated aluminium composite over the control commercially pure aluminium. The creep rate, wear rate and
density are reduced by 75.6, 78.2 and 3.4%, respectively, while the tensile strength,
hardness and melting point are increased by 30.7, 184.8 and 31.4%, respectively.
Except for density, these improvements are quite significant.

From the table, the values of the actual formulated hybrid aluminium composite
and the values predicted from the RSM analysis are reasonably close, with all the
absolute deviations of the predicted from the actual ones not greater than 10%. For
example, the deviations are 5.8, 8.7, 0.7, 2.8, 0.6 and 3.9% for creep rate, wear rate,
density, tensile strength, hardness and melting point, respectively.

Values of the above properties obtained from this research study (i.e. As-cast
values) were compared with existing properties of aluminium alloys used for the
production of engine block (i.e. As-used values). From **Table 5**, out of the six
properties of the engine block examined, the book values of two of the As-used (or
existing) engine block properties (i.e. wear rate and creep rate) were not readily
available in the literature. However, the other four properties (i.e. density, tensile
strength, hardness and melting temperature) were available [27, 28]. The value
of density obtained in this study was found to be within an acceptable range with
the book value. The value of tensile strength was below the range found in literature. However, it is hoped that with further research work, an appreciable tensile
strength value which will fall within the range of values found in the literature

Property	Commercially pure aluminium control	Formulated hybrid aluminium composite	Optimal solution values obtained from RSM (predicted values)	As-used values of existing engine blocks
Creep rate (% elongation/h)	74.04	18.04	19.09	Not available
Wear rate (g/s)	2.61×10^{-4}	0.57×10^{-4}	0.62×10^{-4}	Not available
Density (kg/m^3)	2709.13	2617.39	2598.62	2570–2830
Tensile strength (MPa)	73.99	96.74	94.04	130–280
Hardness (BHN)	98.5	280.49	278.83	80–137
Melting temperature (°C)	685	900	935	645–710

Table 6.
Physical and mechanical properties of the formulated hybrid aluminium composite [14].

would be obtained. The value of the composite material hardness (As-cast value) was found to be far above the range of values of As-used (or existing) engine block. This shows that besides production of IC engine blocks, the developed composite material could also be used for the production of other engineering components where material hardness is of utmost importance. Again, the melting temperature of the developed composite material was found to be far higher than the range of values for aluminium engine blocks (As-used values) as currently used. The developed composite material therefore could withstand the high temperatures that exist within the combustion chamber of IC engines.

3.3 Performance test result of the produced engine block

The fabricated engine block and cylinder liner were assembled into the engine and tested for 1 h. The temperature of the engine block was taken at 1 min intervals. This was repeated for the control cylinder block and liner. The initial temperatures (i.e. before the engine was switched on) of the control engine block and the produced engine block were recorded as 25.6°C (room temperature).

Figure 8 shows the temperature variations for each test. The fabricated engine block had a mean temperature of 151.7°C, while that for the control engine block was 156.9°C giving a deviation of 3.3% from the control value. The lower mean value of the fabricated engine block indicates that it may have a slightly lower heat dissipating ability than the control engine block.

This is expected as it is a fabricated material from aluminium, periwinkle shell and palm kernel shell, the last two being non-metallic materials with lower thermal conductivities. Heywood [29] stated that the combustion temperature within the combustion chamber of the IC engine is about 250°C. Despite this relatively lower heat dissipating ability, the developed hybrid composite material may still be regarded as a good heat conductor because within the first minute, it shot up the engine block temperature from the initial room temperature value of 25.6°C to about 80.7°C, showing an upsurge of 3.2 times the room temperature value. This is comparable to the 3.9 times for the control engine block. This heat conducting phenomenon is good for engine blocks so as to preserve their life span.

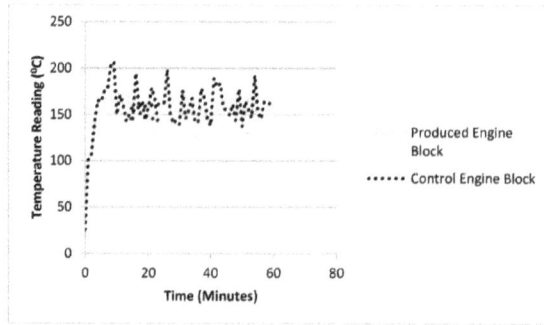

Figure 8.
Temperature reading of produced against control engine blocks.

3.4 Potential benefits of the new aluminium hybrid material

3.4.1 Lightweighting

A gasoline engine rated at 7.1 kW at 5500 rpm generating a maximum torque of 18.0 Nm at 3500 rpm used to power a new lightweight utility vehicle is considered. **Table 7** shows potential weight reductions if this new aluminium hybrid material were to be used as materials for the engine parts indicated, volume for volume. The table shows that:

  i. The weight of the engine block is reduced by 63.03% if the new material were to replace the grey cast iron used.

  ii. The weights of the cylinder head, piston and connecting rod will be reduced by 1.93, 2.13 and 6.37%, respectively, if the new aluminium hybrid material were to replace the aluminium alloys indicated.

  iii. The weight reduction for the engine block, cylinder head, piston and connecting rod totalling 13.501 kg will be 48.37% if the new aluminium hybrid material were used to make these parts.

  iv. The weight reduction for the whole engine weighing 25.75 kg will be 25.36% if the new material were used to make the engine block, cylinder head, piston and connecting rod.

The table also shows the comparison of weight reductions if the conventional aluminium alloy A356 and the new hybrid aluminium material were to be used to make the engine block. As the new material has a slightly lower density, it produces a slightly greater weight reduction of 1.99% over the A356 aluminium alloy. When all the above parts are made with their conventional aluminium alloy materials, the weight reduction will be 2.07% when the new aluminium hybrid material is used for the indicated engine parts.

These results show that lightweighting of the engine and consequently the vehicle powered by it will be achieved if this new aluminium hybrid material is used.

3.4.2 Energy saving

It is envisaged that the use of this new material in manufacturing engines will not only lead to lightweighting but also energy saving. If we consider melting in an

Engine part	Material	Density (kg/m³) [30]	Weight (kg)	Volume (m³)	Weight of equivalent aluminium hybrid material (kg) (density = 2617 kg/m³)	Weight reduction		
						(kg)	(%) (on individual parts)	(%) (on total parts)
Engine block	Grey cast iron	7079	10.25	0.001448 (1448cm³)	3.789	6.461	63.03	—
	Aluminium alloy (A356)	2670	(3.866)	0.001448 (1448cm³)	(3.789)	(0.077)	(1.99)	—
Cylinder head	Aluminium alloy (A356)	2670	3	0.001124 (1124cm³)	2.942	0.058	1.93	—
Piston	Aluminium alloy (A4032)	2690	0.094	0.000035 (35cm³)	0.092	0.002	2.13	—
Connecting rod	Aluminium alloy (A7075)	2803	0.157	0.000056 (56cm³)	0.147	0.010	6.37	—
Total	—	—	13.501 (7.117)	—	6.970 (6.970)	6.531 (0.147)	—	48.37 (2.07)
Whole engine	—	—	25.75	—	—	6.531	—	25.36

Table 7.
Potential lightweighting by the new aluminium hybrid material.

induction furnace, the energy consumption for melting cast iron ranges from 550 to 575 kWh/ton, while that for melting light and solid aluminium scraps ranges from 500 to 625 kWh/ton [31].

Using the average values of the above energy ranges, we have that the:

    i. Cost of energy required to melt the cast iron engine block in **Table 7**
      = ((550 + 575 kWh)/2 × (10.25 kg/1000)) × $0.13/kWh = $0.75.

    ii. Cost of energy required to melt the aluminium hybrid engine block in **Table 7**
      = ((500 + 625 kWh)/2 × (3.789 kg/1000)) × $0.13/kWh = $0.28, where
      $0.13/kWh is the cost of electricity.

The above shows that there could be an energy cost saving of 62.67% in melting which accounts for a significant part of the manufacturing cost, if the new aluminium hybrid material is used in place of cast iron in manufacturing the engine block.

3.5 Further work

This fabricated aluminium composite seems to present an interesting material which requires further work: while its tensile strength is about 53% less than the mean value for the commercially as-used, its hardness and melting points are 159 and 33% more than the mean values for the commercially as-used, respectively. While its tensile strength may further be improved, it could find uses not only for engine blocks but also for other engine parts such as parts of the linkage mechanism in the combustion chamber and other non-engine applications. Also, there is the need to determine its porosity due to these non-metallic combustible materials used as reinforcements.

4. Conclusion

The search for lightweighting materials for automobiles is a continuing process that opens up opportunities for development of new engineering materials. This work which is a part of this process has shown that agro wastes such as palm kernel shell (PKS) and periwinkle shell (PS) may be used as reinforcement materials for metal matrix composites giving good results for fabrication of some IC engine parts such as the engine block. This new material has the potential of lightweighting engines and giving significant energy cost saving in the manufacturing process. Due to the interesting properties it has, further work is necessary to determine additional and/or proper areas of application.

Acknowledgements

The authors wish to thank the Shell Petroleum Development Company of Nigeria Limited (SPDC) for being part sponsors of this work through the Shell Professorial Chair in Lightweight Automobile Engine Development at the Federal University of Petroleum Resources, Effurun, Delta State, Nigeria. The authors also thank the Federal University of Petroleum Resources and the University of Benin, Nigeria, for their support for the project.

Our thanks also go to Dr. P.E. Amiolemhen, Department of Production Engineering, and Dr. E.G. Sadjere, Department of Mechanical Engineering, both of the University of Benin, Nigeria, who were part of the original work.

Author details

Akaehomen O. Akii Ibhadode[1*] and Raphael S. Ebhojiaye[2]

1 Shell Professorial Chair in Lightweight Engine Development, Federal University of Petroleum Resources Effurun, Warri, Nigeria

2 Department of Production Engineering, University of Benin, Benin City, Nigeria

*Address all correspondence to: ibhadode.akii@fupre.edu.ng

IntechOpen

References

[1] Anon. Lightweighting. AutoForm Formimg Reality [Internet]. 2018. Available from: https://www.autoform.com/en/glossary/lightweighting/ [Accessed: October 06, 2018]

[2] Pervaiz M, Panthapulakkal S, Birat KC, Sain M, Tjong J. Emerging trends in automotive lightweighting through novel composite materials. Materials Sciences and Applications. 2016;7:26-38

[3] Isenstadt A and German J (ICCT); Bubna P and Wiseman M (Ricardo Strategic Consulting); Venkatakrishnan U and Abbasov L (SABIC); Guillen P and Moroz N (Detroit Materials); Richman D (Aluminum Association); Kolwich G (FEV), Lightweighting technology development and trends in U.S. passenger vehicles, International Council On Clean Transportation, Washington; 2016

[4] Available from: https://www.shell.com/energy-and-innovation/shell-ecomarathon/europe/results-and-awards/ [Accessed: October 06, 2018]

[5] Pruzc JC, Shoukry SN, William GW, Shoukry MS. Lightweight composite materials for heavy duty vehicles. Final Report: US Department of Energy. Contract No. DE-FC26-08NTO2353; 2013

[6] Available from: https://en.wikipedia.org/wiki/Lightweighting [Accessed: October 06, 2018]

[7] Jacob A. NIST Research Centre Helps Auto Industry Understand Lightweighting Materials [Internet]. 2014. Available from: https://www.materialstoday.com/composite-applications/news/nist-research-centre-helps-auto-industry/ [Accessed: November 06, 2018]

[8] Rajendra K, Babu MVS, Govinda Rao P, Suman KNS. Recent Developments in the Fabrication of Metal Matrix Composites by Stir Casting Route – A Review, International Journal Chemical Science. 2016;14(4):2358-2366

[9] Pal H, Jit N, Tyagi AK, Sidhu S. Metal casting—A general review. Advances in Applied Science Research. 2011;2(5):360-371

[10] Aku SY, Yawas DS, Madakson PB, Amaren SG. Characterization of Periwinkle shell as asbestos-free brake pad materials. The Pacific Journal of Science and Technology. 2012;13(2):57-63

[11] Babafemi AJ, Olusola KO. Influence of curing media on the compressive strength of palm kernel shell (PKS) concrete. IJRRAS. 2012;13(1):180-185

[12] Surappa MK. Aluminum matrix composites: Challenges and opportunities. Sadhana. 2003;28(1 and 2): 319-334

[13] Babalola PO, Bolu CA, Inegbenebor AO, Odunfa KM. Development of aluminum matrix composites: A review. Online International Journal of Engineering and Technology Research. 2014;2:1

[14] Ebhojiaye RS. Development of a hybrid composite material suitable for the production of internal combustion (IC) engine block [PhD thesis]. Benin City, Nigeria: University of Benin; 2018

[15] Kuti OA. Impact of charred palm kernel shell on the calorific value of composite sawdust briquette. Journal of Engineering and Applies Science. 2007;2:62-65

[16] Badmus MAO, Audu TOK, Anyata BU. Removal of lead ion from industrial wastewaters by activated carbon prepared from Periwinkle shell (*Typanotonus fuscatus*). Turkish Journal of Engineering and Environmental Science. 2007;31:251-263

[17] Ebhojiaye RS, Amiolemhen PE, Ibhadode AOA. Development of hybrid composite material of palm kernel shell (PKS) and Periwinkle shell (PS) particles in pure aluminum matrix. In: 2nd International Conference on Innovative and Smart Materials (ICISM 2017); Paris, France: CIUP; December 11-13, 2017

[18] Singh BR, Kumar D, Zaidi MA. A review on stir casting process and parameters. International Journal of Engineering and Management Research. 2017;**7**(3):783-784

[19] Annigeri Veeresh Kumar UKGB. Method of stir casting of aluminum metal matrix composites: A review. Materials Today. 2017;**4**(2, Part A): 1140-1146

[20] Kevin KP, Sijo MT. Effect of stirrer parameter of stir casting on mechanical properties of aluminum silicon carbide composite. International Journal of Modern Engineering Research. 2015;**5**(8):43-49

[21] Ahmad R, Hamidin N, Ali UFM, Abidin CZA. Characterization of bio-oil from palm kernel shell pyrolysis. Journal of Mechanical Engineering and Sciences (JMES). 2014;**7**:1134-1140

[22] Sukumaran K, Pillai SGK, Pillai RM, Kelukutty VS, Pai BC, Satyanarayana KG, et al. The effect of Mg addition on the structure and properties of Al-7Si-10SiCp composites. Journal of Materials Science. 1995;**30**:1469-1472

[23] Soltani S, Azari Khosroshahi R, Mousavian RT, Jiang Z, Fadavi Boostani A, Brabazon D. Stir casting process for manufacture of Al–SiC composites. Rare Metals. 2017;**36**(7):581-590

[24] Jokhio MH, Panhwar MI, Unar MA. Manufacturing of aluminum composite material using stir casting process. Mehran University Research Journal of Engineering & Technology. 2011;**30**(1):53-64

[25] Alaneme KK, Bodunrin MO. Mechanical behaviour of alumina reinforced AA 6063 metal matrix composites developed by two step-stir casting process. ACTA Technica Corvinniensis – Bulletin of Engineering. 2013;**IV**(Fascicule):3

[26] Hashmi J, Looney L, Hashmi MSJ. Metal matrix composites: Production by stir casting method. Journal of Materials Processing Technology, Elsevier B.V, Netherlands. 1999;**92**(93):1-7

[27] Balaji P, Muley AV. Mechanical properties of Al matrix hybrid composites. International Journal of Mechanical and Production Engineering. 2016;**4**(9):1-6

[28] General Motors (GM). Powertrain Casting Development. American Foundry Society (AFS). Available from: https://www.sfsa.org/tutorials/eng_block/GMBlock.pdf. 2004. [Accessed: 2016-05-23]

[29] Heywood JB. Internal Combustion Engine Fundamentals. New York: McGraw-Hill; 1988

[30] Available from: http://www.isibang.ac.in/~library/onlinerz/resources/Enghandbook.pdf [Accessed: October 13, 2018]

[31] Available from: http://www.electroheatinduction.com/how-to-calculate-electricity-cost-for-melting-metal-in-induction-melting-furnace/ [Accessed: October 13, 2018]

www.ingramcontent.com/pod-product-compliance
Lightning Source LLC
Chambersburg PA
CBHW081242190326
41458CB00016B/5877